F# for Scientists

F# for Scientists

Jon Harrop
Flying Frog Consultancy Ltd.

Foreword by Don Syme

WILEY

A JOHN WILEY & SONS, INC., PUBLICATION

Published by John Wiley & Sons, Inc., Hoboken, New Jersey.
Published simultaneously in Canada.

For general information on our other products and services or for technical support, please contact our
Customer Care Department within the United States at (800) 762-2974, outside the United States at
(317) 572-3993 or fax (317) 572-4002.

Wiley also publishes its books in a variety of electronic formats. Some content that appears in print may
not be available in electronic format. For information about Wiley products, visit our web site at
www.wiley.com.

Library of Congress Cataloging-in-Publication Data:

Harrop, Jon D.
    F# for scientists / Jon Harrop.
        p. cm.
    Includes index.
    ISBN 978-0-470-24211-7 (cloth)
1. F# (Computer program language) 2. Functional programming (Computer science) 3. Science—Data
processing. I. Title.
    QA76.73.F163H37 2008
    005.1'14—dc22   2008009567

10 9 8 7 6 5 4 3

To my family

Contents in Brief

CONTENTS

Foreword

Computational science is one of the wonders of the modern world. In almost all areas of science the use of computational techniques is rocketing, and software has moved from being a supporting tool to being a key site where research activities are performed. This has meant a huge increase in the importance of controlling and orchestrating computers as part of the daily routine of a scientific laboratory, from large teams making and running the computers performing global climate simulations to the individual scientist/programmer working alone. Across this spectrum, the productivity of teams and the happiness of scientists depends dramatically on their overall competency as programmers, as well as on their skills as researchers within their field. So, in the last 30 years we have seen the continued rise of that new profession: the *scientific programmer*. A good scientific programmer will carry both epithets with pride, knowing that programming is a key foundation for a successful publication record.

However, programming cultures differ widely, and, over time, gaping divides can emerge that can be to the detriment of all. In this book, Dr. Harrop has taken great steps forward to bridging three very different cultures: *managed code programming*, *scientific programming* and *functional programming*. At a technical level, each has its unique characteristics. Managed code programming, epitomized by .NET and Java, focuses on the productivity of the (primarily commercial) programmer. Scientific programmers focus on high performance computations, data manipulation, numerical computing and visualization. Functional programming focuses on crisp, declarative solutions to problems using compositional techniques. The challenge, then, is to bring these disparate worlds together in a productive way.

The language F#, which Dr. Harrop uses in this book, itself bridges two of these cultures by being a functional language for the .NET platform. F# is an incredibly powerful language: the .NET libraries give a rich and solid foundation of software functionality for many tasks, from routine programming to accessing web services and high performance graphics engines. F# brings an approach to programming that routinely makes even short programs powerful, simple, elegant and correct. However Dr. Harrop has gone a step further, showing how managed code functional programming can revolutionize the art of scientific programming itself by being a powerful workhorse tool that unifies and simplifies many of the tasks scientific programmers face.

But what of the future? The next 20 years will see great changes in scientific programming. It is customary to mention the ever-increasing challenges of parallel, concurrent, distributed and reactive programming. It is widely expected that future micro-processors will use ever-increasing transistor counts to host multiple processing cores, rather than more sophisticated microprocessor designs. If computations can be parallelized and distributed on commodity hardware then the computing resources that can be brought can be massively increased. It is well known that successful concurrent and distributed computing requires a combination of intelligent algorithm design, competent programming, and core components that abstract some details of concurrent execution, e.g. databases and task execution libraries. This needs a language that can interoperate with key technologies such as databases, and parallelism engines. Furthermore, the ability to declaratively and crisply describe solutions to concurrent programming problems is essential, and F# is admirably suited to this task.

The future, will, however, bring other challenges as well. Truly massive amounts of data are now being generated by scientific experiments. Web-based programming will become more and more routine for scientific teams: a good web application can revolutionize a scientific field. Shared databases will soon be used in almost every scientific field, and programmatic access to these will be essential. F# lends itself to these challenges: for example, it is relatively easy to perform sophisticated and high-performance analysis of these data sources by bringing them under the static type discipline of functional programming, as shown by some of the samples in this book.

You will learn much about both programming and science through this book. Dr. Harrop has chosen the style of F# programming most suited to the individual scientist: crisp, succinct and efficient, with a discursive presentation style reminiscent of Mathematica. It has been a pleasure to read, and we trust it will launch you on a long and productive career as a managed code, functional scientific programmer.

Preface

The face of scientific computing has changed. Computational scientists are no longer writing their programs in Fortran and competing for time on supercomputers. Scientists are now streamlining their research by choosing more expressive programming languages, parallel processing on desktop machines and exploiting the wealth of scientific information distributed across the internet.

The landscape of programming languages saw a punctuation in its evolution at the end of the 20th century, marked by the advent of a new breed of languages. These new languages incorporate a multitude of features that are all designed to serve a single purpose: to make life easier. Modern programming languages offer so much more expressive power than traditional languages that they even open up new avenues of scientific research that were simply intractable before.

The next few years will usher in a new era of computing, where parallelism becomes ubiquitous. Few approaches to programming will survive this transition, and functional programming is one of them.

Seamlessly interoperating with computers across the world is of pivotal importance not only because of the breadth of information now available on-line but also because this is the only practicable way to interrogate the enormous amount of data available. The amount of genomic and proteinomic data published every year continues to grow exponentially, as each generation of technology fuels the next.

Only one mainstream programming language combines awesome expressive power, interoperability and performance: F#. This book introduces all of the aspects of the F# programming language needed by a working scientist, emphasizing aspects not covered by existing literature. Consequently, this book is the ideal complement to a detailed overview of the language itself, such as the F# manual or the book Expert F#[25].

Chapters 1–5 cover the most important aspects of F# programming needed to start developing useful F# programs. Chapter 6 ossifies this knowledge with a variety of enlightening and yet simple examples. Chapters 7–11 cover advanced topics including real-time visualization, interoperability and parallel computing. Chapter 12 concludes the book with a suite of complete working programs relevant to scientific computing.

The source code from this book is available from the following website:

`http://www.ffconsultancy.com/products/fsharp_for_scientists/`

J. D. HARROP

Cambridge, UK
June, 2008

Acknowledgments

I would like the thank Don Syme, the creator of F#, for pioneering research into programming languages and for thrusting the incredibly powerful ML family of languages into the limelight of mainstream software development.

Xavier Leroy, everyone at projet Cristal and the Debian package maintainers for making OCaml so practically useful.

Stephen Elliott and Sergei Taraskin and their group at the University of Cambridge for teaching me how to be a research scientist and letting me pursue crazy ideas when I should have been working.

Ioannis Baltopoulos and Enrique Nell for proofreading this book and giving essential feedback.

J. D. H.

List of Figures

List of Tables

Acronyms

ADO	Active Data Objects
ASP	Active Server Pages
AST	Abstract-syntax tree
BNF	Backus-Naur form
CAML	Categorical Abstract Machine Language
FFT	Fast Fourier Transform
FFTW	Fastest Fourier Transform in the West
GOE	Gaussian Orthogonal Ensemble
HOF	Higher-Order Function
IDE	Integrated Development Environment
INRIA	Institut National de Recherche en Informatique et en Automatique
IO	Input and Output
LCF	Logic of Computable Functions
ML	Meta-Language

OCaml	Objective CAML
OO	Object-Oriented
OOP	Object-Oriented Programming
OpenGL	Open Graphics Library
RPC	Remote Procedure Call
SOAP	Simple Object Access Protocol
UDDI	Universal Description, Discovery and Integration
VM	Virtual Machine
VS	Visual Studio
WSDL	Web Service Definition Language
XML	eXtensible Markup Language
XSLT	eXtensible Stylesheet Language Transformations

CHAPTER 1

INTRODUCTION

For the first time in history, and thanks to the exponential growth rate of computing power, an increasing number of scientists are finding that more time is spent creating, rather than executing, working programs. Indeed, much effort is spent writing small programs to automate otherwise tedious forms of analysis. In the future, this imbalance will doubtless be addressed by the adoption and teaching of more efficient programming techniques. An important step in this direction is the use of higher-level programming languages, such as F#, in place of more conventional languages for scientific programming such as Fortran, C, C++ and even Java and C#.

In this chapter, we shall begin by laying down some guidelines for good programming which are applicable in any language before briefly reviewing the history of the F# language and outlining some of the features of the language which enforce some of these guidelines and other features which allow the remaining guidelines to be met. As we shall see, these aspects of the design of F# greatly improve reliability and development speed. Coupled with the fact that a freely available, efficient compiler already exists for this language, no wonder F# is already being adopted by scientists of all disciplines.

1.1   PROGRAMMING GUIDELINES

Some generic guidelines can be productively adhered to when programming in any language:

Correctness over performance   Programs should be written correctly first and optimized last.

Factor programs   Complicated or common operations should be factored out into separate functions or objects.

Interfaces   Abstract interfaces should be designed and concrete implementations should be coded to these interfaces.

Avoid magic numbers   Numeric constants should be defined once and referred back to, rather than explicitly "hard-coding" their value multiple times at different places in a program.

Following these guidelines is the first step towards reusable programs.

1.2   A BRIEF HISTORY OF F#

The first version of ML (Meta Language) was developed at Edinburgh University in the 1970's as a language designed to efficiently represent and manipulate other languages. The original ML language was pioneered by Robin Milner for the *Logic of Computable Functions* (**LCF**) theorem prover. The original ML, and its derivatives, were designed to stretch theoretical computer science to the limit, yielding remarkably robust and concise programming languages without sacrificing the performance of low-level languages.

The *Categorical Abstract Machine Language* (**CAML**) was the acronym originally used to describe what is now known as the Caml family of languages, a dialect of ML that was designed and implemented by Gérard Huet at the *Institut National de Recherche en Informatique et en Automatique* (**INRIA**) in France, until 1994. Since then, development has continued as part of *projet Cristal*, now led by Xavier Leroy. *Objective Caml* (**OCaml**) is the current flagship language of projet Cristal. The OCaml programming language is one of the foremost high-performance and high-level programming languages used by scientists on the Linux and Mac OS X platforms [11].

Don Syme at Microsoft Research Cambridge has meticulously engineered the F# language for .NET, drawing heavily upon the success of the CAML family of languages. The F# language combines the remarkable brevity and robustness of the Caml family of languages with .NET interoperability, facilitating seamless integration of F# programs with any other programs written in .NET languages. Moreover, F# is the first mainstream language to implement some important features such as active patterns and asynchronous programming constructs.

1.3   BENEFITS OF F#

Before delving into the syntax of the language itself, we shall list the main, advantageous features offered by the F# language:

Safety F# programs are thoroughly checked prior to execution such that they are proven to be entirely safe to run, e.g. a compiled F# program cannot cause an access violation.

Functional Functions may be nested, passed as arguments to other functions and stored in data structures as values.

Strongly typed The types of all values are checked during compilation to ensure that they are well defined and validly used.

Statically typed Any typing errors in a program are picked up at compile-time by the compiler, instead of at run-time as in many other languages.

Type inference The types of values are automatically inferred during compilation by the context in which they occur. Therefore, the types of variables and functions in F# code rarely need to be specified explicitly, dramatically reducing source code size. Clarity is regaining by displaying inferred type information in the integrated development environment (**IDE**).

Generics Functions are automatically generalized by the F# compiler, greatly simplifying the writing of reusable functions.

Pattern matching Values, particularly the contents of data structures, can be matched against arbitrarily-complicated patterns in order to determine the appropriate course of action.

Modules and objects Programs can be structured by grouping their data structures and related functions into modules and objects.

Separate compilation Source files can be compiled separately into object files that are then linked together to form an executable or library. When linking, object files are automatically type checked and optimized before the final executable is created.

Interoperability F# programs can call and be called from programs written in other Microsoft .NET languages (e.g. C#), native code libraries and over the internet.

1.4   INTRODUCING F#

F# programs are typically written in Microsoft Visual Studio and can be executed either following a complete build or incrementally from the F# interactive mode. Throughout this book we shall present code snippets in the form seen using the F# interactive mode, with code input following the prompt:

```
>
```

Setup and use of the interactive mode is covered in more detail in chapter 2. Throughout this book, we assume the use of the #light syntax option, which requires the following command to be evaluated before any of the code examples:

```
> #light;;
```

Before we consider the features offered by F#, a brief overview of the syntax of the language is instructive, so that we can provide actual code examples later. Other books give more systematic, thorough and formal introductions to the whole of the F# language [25, 22].

1.4.1 Language overview

In this section we shall evolve the notions of values, types, variables, functions, simple containers (lists and arrays) and program flow control. These notions will then be used to introduce more advanced features in the later sections of this chapter.

When presented with a block of code, even the most seasoned and fluent programmer will not be able to infer the purpose of the code. Consequently, programs should contain additional descriptions written in plain English, known as *comments*. In F#, comments are enclosed between (* and *) or after // or /// on a single line. Comments appearing after a /// are known as *autodoc comments* and Visual Studio interprets them as official documentation according to standard .NET coding guidelines.

Comments may be nested, i.e. (* (* ... *) *) is a valid comment and comments are treated as whitespace, i.e. a(* ... *)b is understood to mean a b rather than ab.

Just as numbers are the members of sets such as the integers ($\in \mathbb{Z}$), reals ($\in \mathbb{R}$), complexes ($\in \mathbb{C}$) and so on, so *values* in programs are members of sets. These sets are known as *types*.

1.4.1.1 *Basic types*   Fundamentally, languages provide basic types and, often, allow more sophisticated types to be defined in terms of the basic types. F# provides a number of built-in types, such as unit, int, float, char, string and bool. We shall examine these built-in types before discussing the compound *tuple*, *record* and *variant* (also known as *discriminated union*) types.

Only one value is of type unit and this value is written () and, therefore, conveys no information. This is used to implement functions that require no input or expressions that return no value. For example, a new line can be printed by calling the print_newline function:

```
> print_newline ();;

val it : unit = ()
```

This function requires no input, so it accepts a single argument () of the type unit, and returns the value () of type unit.

Integers are written -2, -1, 0, 1 and 2. Floating-point numbers are written -2.0, -1.0, -0.5, 0.0, 0.5, 1.0 and 2.0. Note that a zero fractional part may be omitted, so 3.0 may be written 3., but we choose the more verbose format for purely esthetic reasons. For example:

```
> 3;;
val it : int = 3
> 5.0;;
val it : float = 5.0
```

Arithmetic can be performed using the conventional +, -, *, / and % binary infix[1] operators over many arithmetic types including int and float.

For example, the following expression is evaluated according to usual mathematical convention regarding operator precedence, with multiplication taking precedence over addition:

```
> 1 * 2 + 2 * 3;;
val it : int = 8
```

The same operators can be used for floating point arithmetic:

```
> 1.0 * 2.0 + 2.0 * 3.0;;
val it : float = 8.0
```

Defining new operators and overloading existing operators is discussed later, in section 2.4.1.3. Conversion functions or *type casts* are used to perform arithmetic with mixed types, e.g. the float function converts numeric types to the float type.

However, the types of the two arguments to these operators must be the same, so * cannot be used to multiply an int by a float:

```
> 2 * 2.0;;
Error: FS0001: This expression has type float but is
here used with type int
```

Explicitly converting the value of type float to a value of type int using the built-in function int results in a valid expression that the interactive session will execute:

```
> 2 * int 2.0;;
val it : int = 4
```

In most programs, arithmetic is typically performed using a single number representation (e.g. either int or float) and conversions between representations are, therefore, comparatively rare. Thus, the overhead of having to apply functions to explicitly convert between types is a small price to pay for the added robustness that results from more thorough type checking.

---

[1]An infix function is a function that appears between its arguments rather than before them. For example, the arguments i and j of the conventional addition operator $+$ appear on either side: $i + j$.

Single characters (of type char) are written in single quotes, e.g. `'a'`, that may also be written using a 3-digit decimal code, e.g. `'\097'`.

Strings are written in double quotes, e.g. `"Hello World!"`. Characters in a string of length n may be extracted using the notation `s.[i]` for $i \in \{0 \dots n-1\}$. For example, the fifth character in this string is "o":

```
> "Hello world!".[4];;
val it : char = 'o'
```

Strings are immutable in F#, i.e. the characters in a string cannot be altered once the string is created. The `char array` and `byte array` types may be used as mutable strings.

A pair of strings may be concatenated using the overloaded + operator:

```
> "Hello " + "world!";;
val it : string = "Hello world!"
```

Booleans are either `true` or `false`. Booleans are created by the usual comparison functions =, <> (not equal to), <, >, <=, >=. These functions are polymorphic, meaning they may be applied to pairs of values of the same type for any type. The usual, short-circuit-evaluated[2] logical comparisons && and || are also present. For example, the following expression tests that one is less than three and 2.5 is less than 2.7:

```
> 1 < 3 && 2.5 < 2.7;;
val it : bool = true
```

Values may be assigned, or *bound*, to names. As F# is a functional language, these values may be expressions that map values to values — functions. We shall now examine the binding of values and expressions to variable and function names.

1.4.1.2 Variables and functions   Variables and functions are both defined using the `let` construct. For example, the following defines a variable called a to have the value 2:

```
> let a = 2;;
val a : int
```

Note that the language automatically infers types. In this case, a has been inferred to be of type `int`.

Definitions using `let` can be defined locally using the syntax:

`let` *var = expr*$_1$ `in`
expr$_2$

This evaluates *expr*$_1$ and binds the result to the variable *var* before evaluating *expr*$_2$. For example, the following evaluates a^2 in the context $a = 3$, giving 9:

---

[2]Short-circuit evaluation refers to the premature escaping of a sequence of operations (in this case, boolean comparisons). For example, the expression `false && ` *expr* need not evaluate *expr* as the result of the whole expression is necessarily `false` due to the preceding `false`.

```
> let a = 3 in
  a * a;;
val it : int = 9
```

Note that the value 3 bound to the variable a in this example was local to the expression a * a and, therefore, the global definition of a is still 2:

```
> a;;
val it : int = 2
```

More recent definitions shadow previous definitions. For example, the following supersedes a definition $a = 5$ with $a = a \times a$ in order to calculate $5^4 = 625$:

```
> let a = 5;;
val a : int
> let a = a * a;;
val a : int
> a * a;;
val it : int = 625
```

Note that many of the keywords at the ends of lines (such as the in keyword) may be omitted when using the #light syntax option. This simplifies F# code and makes it easier to read. More importantly, nested lines of code are written in the same style and may be evaluated directly in a running F# interactive session. This is discussed in chapter 2.

As F# is a functional language, values can be functions and variables can be bound to them in exactly the same way as we have just seen. Specifically, function definitions include a list of arguments between the name of the function and the = in the let construct. For example, a function called sqr that accepts an argument n and returns n * n may be defined as:

```
> let sqr n = n * n;;
val sqr : int -> int
```

Type inference for arithmetic operators defaults to int. In this case, the use of the overloaded multiply * results in F# inferring the type of sqr to be int -> int, i.e. the sqr function accepts a value of type int and returns a value of type int.

The function sqr may then be applied to an int as:

```
> sqr 5;;
val it : int = 25
```

In order to write a function to square a float, it is necessary to override this default type inference. This can be done by explicitly annotating the type. Types may be constrained by specifying types in a definition using the syntax (*expr* : *type*). For example, specifying the type of the argument alters the type of the whole function:

```
> let sqr (x : float) = x * x;;
val sqr : float -> float
```

The return type of a function can also be constrained using a similar syntax, having the same result in this case:

```
> let sqr x : float = x * x;;
val sqr : float -> float
```

A variation on the `let` binding called a `use` binding is used to automatically dispose a value at the end of the scope of the `use` binding. This is particularly useful when handling file streams (discussed in chapter 5) because the file is guaranteed to be closed.

Typically, more sophisticated computations require the use of more complicated types. We shall now examine the three simplest ways by which more complicated types may be constructed.

1.4.1.3 Product types: tuples and records

Tuples are the simplest form of compound types, containing a fixed number of values which may be of different types. The type of a tuple is written analogously to conventional set-theoretic style, using * to denote the cartesian product between the sets of possible values for each type. For example, a tuple of three integers, conventionally denoted by the triple $(i, j, k) \in \mathbb{Z} \times \mathbb{Z} \times \mathbb{Z}$, can be represented by values (i, j, k) of the type int * int * int. When written, tuple values are comma-separated and often enclosed in parentheses. For example, the following tuple contains three different values of type int:

```
> (1, 2, 3);;
val it : int * int * int = (1, 2, 3)
```

At this point, it is instructive to introduce some nomenclature: A tuple containing n values is described as an n-tuple, e.g. the tuple (1, 2, 3) is a 3-tuple. The value n is said to be the *arity* of the tuple.

Records are essentially tuples with named components, known as *fields*. Records and, in particular, the names of their fields must be defined using a `type` construct before they can be used. When defined, record fields are written *name* : *type* where *name* is the name of the field (which must start with a lower-case letter) and *type* is the type of values in that field, and are semicolon-separated and enclosed in curly braces. For example, a record containing the x and y components of a 2D vector could be defined as:

```
> type vec2 = { x : float; y : float };;
type vec2 = { x:float; y:float }
```

A value of this type representing the zero vector can then be defined using:

```
> let zero = { x = 0.0; y = 0.0 };;
val zero : vec2
```

Note that the use of a record with fields x and y allowed F# to infer the type of zero as vec2.

Whereas the tuples are order-dependent, i.e. $(1, 2) \neq (2, 1)$, the named fields of a record may appear in any order, i.e. $\{x = 1; y = 2\} \equiv \{y = 2; x = 1\}$. Thus, we could, equivalently, have provided the x and y fields in reverse order:

```
> let zero = { y = 0.0; x = 0.0 };;
val zero : vec2
```

The fields in this record can be extracted individually using the notation *record* .*field* where *record* is the name of the record and *field* is the name of the field within that record. For example, the x field in the variable zero is 0:

```
> zero.x;;
val it : float = 0.0
```

Also, a shorthand with notation exists for the creation of a new record from an existing record with some of the fields replaced. This is particularly useful when records contain many fields. For example, the record $\{x=1.0; y=0.0\}$ may be obtained by replacing the field x in the variable zero with 1:

```
> let x_axis = { zero with x = 1.0 };;
val x_axis : vec2
> x_axis;;
val it : vec2 = {x = 1.0; y = 0.0}
```

Like many operations in F#, the with notation leaves the original record unaltered, creating a new record instead.

1.4.1.4 Sum types: variants   The types of values stored in tuples and records are known at compile-time. The F# compiler enforces the correct use of these types at compile-time. However, this is too restrictive in many circumstances. These requirements can be slightly relaxed by allowing a type to be defined which can acquire one of several possible types at run-time. These are known as *variant types*.

Variant types are defined using the type construct with the possible constituent types referred to by *constructors* (the names of which must begin with upper-case letters) separated by the | character. For example, a variant type named button that may adopt the values On or Off may be written:

```
> type button =
    | On
    | Off;;
type button = On | Off
```

The constructors On and Off may then be used as values of type button:

```
> On;;
val it : button = On
> Off;;
val it : button = Off
```

In this case, the constructors On and Off convey no information in themselves (i.e. like the type unit, On and Off do not carry data) but the choice of On or Off does convey information. Note that both expressions were correctly inferred to be of type button.

More usefully, constructors may take arguments, allowing them to convey information by carrying data. The arguments are defined using of and are written in the same form as that of a tuple. For example, a replacement button type which provides an On constructor accepting two arguments (and int and a string) may be written:

```
> type button =
    | On of int * string
    | Off;;
type button = On of int * string | Off
```

The On constructor may then be used to create values of type button by appending the argument in the style of a tuple:

```
> On (1, "mine");;
val it : button = On (1, "mine")
> On (2, "hers");;
val it : button = On (2, "hers")
> Off;;
val it : button = Off
```

Types can also be defined recursively, which is very useful when defining more sophisticated data structures, such as trees. For example, a binary tree contains either zero or two binary trees and can be defined as:

```
> type binary_tree =
    | Leaf
    | Node of binary_tree * binary_tree;;
type binary_tree =
  | Leaf
  | Node of binary_tree * binary_tree
```

A value of type binary_tree may be written in terms of these constructors:

```
> Node (Node (Leaf, Leaf), Leaf);;
val it : binary_tree = Node (Node (Leaf, Leaf), Leaf)
```

Of course, we could also place data in the nodes to make a more useful data structure. This line of thinking will be pursued in chapter 3. In the meantime, let us consider two special data structures which have notations built into the language.

1.4.1.5 Generics The automatic generalization of function definitions to their most generic form is one of the critical benefits offered by the F# language. In order to exploit such genericity it is essential to be able to parameterize tuple, record and variant types over type variables. A type variable is simply the type theory equivalent of a variable in mathematics. In any given type expression, a type variable denotes any type and a concrete type may be substituted accordingly. Type variables are written ' a, ' b and so on.

For example, the following function definition handles a 2-tuple (a pair) but the type of the two elements of the pair are not known and, consequently, the F# compiler automatically generalizes the function to apply to pairs of any two types denoted ' a and ' b, respectively:

```
> let swap(a, b) = b, a;;
val swap : 'a * 'b -> 'b * 'a
```

Note that the behaviour of this swap function, to swap the elements of a pair, is reflected in its type because the type variables appear in reverse order in the return value. As a programmer grows accustomed to the implications of inferred types, the types of expressions and function definitions come to convey a significant amount of information. Moreover, the type information printed explicitly following an interactive definition or expression in an F# interactive session (as shown here) is made available directly from the source code in an IDE such as Visual Studio. This is described in more detail in chapter 2.

So the type of a generic pair is written ' a * ' b, the type of a generic record t is written (' a, ' b) t. For example, the previous record type vec may be parameterized over a generic field type ' a. This is defined and used as follows:

```
> type 'a vec = { x: 'a; y: 'a };;
type 'a vec = { x: 'a; y: 'a }
> { x = 3.0; y = 4.0 };;
val it : float vec = { x = 3.0; y = 4.0 }
```

Note that the parameterized record type is referred to generically as ' a vec and specifically in this case as float vec because the elements are of the type float.

Generic variant types are written in an equivalent notation. For example, the following defines and uses a generic variant type called ' a option:

```
> type 'a option =
    | None
    | Some of 'a;;
type 'a option = None | Some of 'a
> None;;
val it : 'a option = None
> Some 3;;
val it : int option = Some 3
```

This type is actually so useful that it is provided by the F# standard library. Many of the built in data structures, including lists and arrays, are parameterized over the type of elements they contain and generic functions and types are used extensively in the remainder of this book.

In the context of generic classes and generic .NET data structures, a generic type ' a t is often written equivalently as t<' a>. This alternative syntax arises in section 1.4.4 in the context of sequence expressions.

1.4.1.6 Lists and arrays   Lists are written [a; b; c] and arrays are written [|a; b; c|]. As we shall see in chapter 3, lists and arrays have different merits.

Following the notation for generic types, the types of lists and arrays of integers are written int list and int array, respectively:

```
> [1; 2; 3];;
val it : int list = [1; 2; 3]
> [|1; 2; 3|];;
val it : int array = [|1; 2; 3|]
```

In the case of lists, the infix cons operator :: provides a simple way to prepend an element to the front of a list. For example, prepending 1 onto the list [2; 3] gives the list [1; 2; 3]:

```
> 1 :: [2; 3];;
val it : int list = [1; 2; 3]
```

In the case of arrays, the notation *array*.[*i*] may be used to extract the $i + 1^{\text{th}}$ element. For example, [|3; 5; 7|].[1] gives the second element 5:

```
> [|3; 5; 7|].[1];;
val it : int = 5
```

Also, a short-hand notation can be used to represent lists or arrays of tuples by omitting unnecessary parentheses. For example, [(a, b); (c, d)] may be written [a, b; c, d].

The use and properties of lists, arrays and several other data structures will be discussed in chapter 3. In the mean time, we shall examine programming constructs which allow more interesting computations to be performed.

1.4.1.7 The if expression   Like many other programming languages, F# provides an if construct which allows a boolean "predicate" expression to determine which of two expressions is evaluated and returned, as well as a special if construct which optionally evaluates an expression of type unit:

if *expr*₁ then *expr*₂
if *expr*₁ then *expr*₂ else *expr*₃

In both cases, *expr*₁ must evaluate to a value of type bool. In the former case, *expr*₂ is expected to evaluate to the value of type unit. In the latter case, both *expr*₂ and *expr*₃ must evaluate to values of the same type.

The former evaluates the boolean expression *expr*₁ and, only if the result is true, evaluates the expression *expr*₂. Thus, the former is equivalent to:

if *expr*₁ then *expr*₂ else ()

The latter similarly evaluates *expr*₁ but returning the result of either *expr*₂, if *expr*₁ evaluated to true, or of *expr*₃ otherwise.

For example, the following function prints "Less than three" if the given argument is less than three:

```
> let f x =
    if x < 3 then
      print_endline "Less than three";;
val f : int -> unit
> f 5;;
val it : unit = ()
> f 1;;
Less than three
val it : unit = ()
```

The following function returns the string "Less" if the argument is less than 3 and "Greater" otherwise:

```
> let f x =
    if x < 3 then
      "Less"
    else
      "Greater";;
val f : int -> string
> f 1;;
val it : string = "Less"
> f 5;;
val it : string = "Greater"
```

The if expression is significantly less common in F# than many other languages because a much more powerful form of run-time dispatch is provided by pattern matching, which will be introduced in section 1.4.2.

1.4.1.8 *More about functions* Functions can also be defined anonymously, known as λ-abstraction in computer science. For example, the following defines a function $f(x) = x \times x$ which has a type representing[3] $f : \mathbb{Z} \to \mathbb{Z}$:

```
> fun x -> x * x;;
val it : int -> int = <fun:clo@0_3>
```

This is an anonymous equivalent to the sqr function defined earlier. The type of this expression is also inferred to be int -> int. This anonymous function may be applied as if it were the name of a conventional function. For example, applying the function f to the value 2 gives $2 \times 2 = 4$:

```
> (fun x -> x * x) 2;;
val : int = 4
```

Consequently, we could have defined the sqr function equivalently as:

```
> let sqr = fun x -> x * x;;
```

---

[3] We say "representing" because the F# type int is, in fact, a finite subset of $\mathbb{Z}$, as we shall see in chapter 4.

```
val sqr : int -> int
```

Once defined, this version of the sqr function is indistinguishable from the original.

The let ... in construct allows definitions to be nested, including function definitions. For example, the following function ipow3 raises a given int to the power three using a sqr function nested within the body of the ipow3 function:

```
> let ipow3 x =
    let sqr x = x * x
    x * sqr x;;
val ipow3 : int -> int
```

Note that the #light syntax option allowed us to omit the in keyword from the inner let binding, and that the function application sqr x takes precedence over the multiplication.

The let construct may also be used to define the elements of a tuple simultaneously. For example, the following defines two variables, a and b, simultaneously:

```
> let a, b = 3, 4;;
val a : int
val b : int
```

This is particularly useful when factoring code. For example, the following definition of the ipow4 function contains an implementation of the sqr function which is identical to that in our previous definition of the ipow3 function:

```
> let ipow4 x =
    let sqr x = x * x
    sqr(sqr x);;
val ipow4 : int -> int
```

Just as common subexpressions can be factored out of a mathematical expression, so the ipow3 and ipow4 functions can be factored by sharing a common sqr function and returning the ipow3 and ipow4 functions simultaneously in a 2-tuple:

```
> let ipow3, ipow4 =
    let sqr x = x * x
    (fun x -> x * sqr x), (fun x -> sqr(sqr x));;
val ipow3 : int -> int
val ipow4 : int -> int
```

Factoring code is an important way to keep programs manageable. In particular, programs can be factored much more aggressively through the use of higher-order functions (**HOF**s) — something that can be done in F# but not Java, C++ or Fortran. We shall discuss such factoring of F# programs as a means of code structuring in chapter 2. In the meantime, we shall examine *recursive* functions, which perform computations by applying themselves.

As we have already seen, variable names in let definitions refer to their previously defined values. This default behaviour can be overridden using the rec keyword, which allows a variable definition to refer to itself. This is necessary to define a recursive function[4]. For example, the following implementation of the `ipow` function, which computes n^m for $n, m \geq 0 \in \mathbb{Z}$, calls itself recursively with smaller m to build up the result until the base-case $n^0 = 1$ is reached:

```
> let rec ipow n m =
    if m = 0 then 1 else
      n * ipow n (m - 1);;
val ipow : int -> int -> int
```

For example, $2^{16} = 65,536$:

```
> ipow 2 16;;
val it : int = 65536
```

Recursion is an essential construct in functional programming and will be discussed in more detail in section 1.6.

The programming constructs described so far may already be used to write some interesting functions, using recursion to act upon values of non-trivial types. However, one important piece of functionality is still missing: the ability to dissect variant types, dispatching according to constructor and extracting any data contained in them. Pattern matching is an incredibly powerful core construct in F# that provides exactly this functionality.

1.4.2 Pattern matching

As a program is executed, it is quite often necessary to choose the future course of action based upon the value of a previously computed result. As we have already seen, a two-way choice can be implemented using the `if` construct. However, the ability to choose from several different possible actions is often desirable. Although such cases can be reduced to a series of `if` tests, languages typically provide a more general construct to compare a result with several different possibilities more succinctly, more clearly and sometimes more efficiently than manually-nested `if`s. In Fortran, this is the `SELECT CASE` construct. In C and C++, it is the `switch case` construct.

Unlike conventional languages, F# allows the value of a previous result to be compared against various patterns - *pattern matching*. As we shall see, this approach is considerably more powerful and even more efficient than the conventional approaches.

The most common pattern matching construct in F# is in the `match` ... `with` ... expression:

```
match expr with
| pattern₁ -> expr₁
| pattern₂ -> expr₂
```

---

[4]A recursive function is a function that calls itself, possibly via other functions.

```
| pattern₃  -> expr₃
| ...
| patternₙ  -> exprₙ
```

This evaluates *expr* and compares the resulting value firstly with *pattern*$_1$ then with *pattern*$_2$ and so on, until a pattern is found to match the value of *expr*, in which case the corresponding expression *expr*$_n$ is evaluated and returned.

Patterns may reflect arbitrary data structures (tuples, records, variant types, lists and arrays) that are to be matched verbatim and, in particular, the cons operator : : may be used in a pattern to decapitate a list. Also, the pattern _ matches any value without assigning a name to it. This is useful for clarifying that part of a pattern is not referred to in the corresponding expression.

For example, the following function f compares its argument i against three patterns, returning the expression of type string corresponding to the first pattern that matches:

```
> let f i =
    match i with
    | 0 -> "Zero"
    | 3 -> "Three"
    | _ -> "Neither zero nor three";;
val f : int -> string
```

Applying this function to some expressions of type int demonstrates the most basic functionality of the match construct:

```
> f 0;;
val it : string = "Zero"
> f 1;;
val it : string = "Neither zero nor three"
> f (1 + 2);;
val it : string = "Three"
```

As pattern matching is such a fundamental concept in F# programming, we shall provide several more examples using pattern matching in this section.

A function is_empty_list which examines a given list and returns true if the list is empty and false if the list contains any elements, may be written without pattern matching by simply testing equality with the empty list:

```
> let is_empty_list list =
    list = [];;
val is_empty_list : 'a list -> bool
```

Note that the clean design of the F# language allows the identifier list to be used to refer to both a variable name and a type without conflict.

Using pattern matching, this example may be written using the match ... with ... construct as:

```
> let is_empty_list list =
```

```
    match list with
    | [] -> true
    | _::_ -> false;;
val is_empty_list : 'a list -> bool
```

Note the use of the anonymous _ pattern to match any value, in this case accounting for all other possibilities.

The is_empty_list function can also be written using the function ... construct, used to create one-argument λ-functions which are pattern matched over their argument:

```
> let is_empty_list = function
    | [] -> true
    | _::_ -> false;;
val is_empty_list : 'a list -> bool
```

In general, functions that pattern match over their last argument may be rewritten more succinctly using function.

1.4.2.1 Variables in patterns   Variables that are named in the pattern on the left hand side of a match case are bound to the corresponding parts of the value being matched when evaluating the corresponding expression on the right hand side of the match case. This allows parts of a data structure to be used in the resulting computation and, in particular, this is the only way to extract the values of the arguments of a variant type constructor.

The following function f tries to extract the argument of the Some constructor of the built in option type, returning a default value of 0 if the given value is None:

```
> let f = function
    | None -> 0
    | Some x -> x;;
val f : int option -> int
```

Note that the default value of 0 returned by the first match case of this pattern match led type inference to determine that the argument to the f function must be of the type int option.

For example, applying this function the values None and Some 3 gives the results 0 and 3 as expected:

```
> f None;;
val it : int = 0
> f(Some 3);;
val it : int = 3
```

In the latter case, the second pattern is matched and the variable name x appearing in the pattern is bound to the corresponding value in the data structure, which is 3 in this case. The second match case simply returns the value bound to x, returning 3 in this case.

The ability to deconstruct the value of a variant type into its constituent parts is the single most important use of pattern matching.

1.4.2.2 Named subpatterns   Part of a data structure can be bound to a variable by giving the variable name in the pattern. Occasionally, it is also useful to be able to bind part of a data structure matched by a sub pattern.

The `as` construct provides this functionality, allowing the value corresponding to a matched subpattern to be bound to a variable.

For example, the following recursive function returns a list of all adjacent pairs from the given list:

```
> let rec pairs = function
    | h1::(h2::_ as t) -> (h1, h2) :: pairs t
    | _ -> [];;
val pairs : 'a list -> ('a * 'a) list
```

In this case, the first two elements from the input are named `h1` and `h2` and the tail list after `h1` is named `t`, i.e. `h2` is the head of the tail list `t`.

Applying the `pairs` function to an example list returns the pairs of adjacent elements as a list of 2-tuples:

```
> pairs [1; 2; 3; 4; 5];;
val it : (int * int) list =
  [(1, 2); (2, 3); (3, 4); (4, 5)]
```

Named subpatterns are used in some of the later examples in this book.

1.4.2.3 Guarded patterns   Patterns may also have arbitrary tests associated with them, written using the `when` construct. Such patterns are referred to as *guarded patterns* and are only allowed to match when the associated boolean expression (the *guard*) evaluates to true.

For example, the following recursive function filters out only the non-negative numbers from a list:

```
> let rec positive = function
    | [] -> []
    | h::t when h < 0 -> positive t
    | h::t -> h::positive t;;
val positive : int list -> int list
```

Applying this function to a list containing positive and negative numbers results in a list with the negative numbers removed

```
> positive [-3; 1; -1; 4];;
val it : int list = [1; 4]
```

Although guarded patterns undermine some of the static checking of pattern matches that the F# compiler can perform, they can be used to good effect in a variety of circumstances.

1.4.2.4 Or patterns   In many cases it is useful for several different patterns to be combined into a single pattern that is matched when any of the alternatives matches. Such patterns are known as *or-patterns* and use the syntax:

pattern$_1$ | *pattern*$_2$

Or patterns must bind the same sets of variables.

For example, the following function returns `true` when its argument is in $\{-1, 0, 1\}$ and `false` otherwise:

```
> let is_sign = function
    | -1 | 0 | 1 -> true
    | _ -> false;;
val is_sign : int -> bool
```

The sophistication provided by pattern matching may be misused. Fortunately, the F# compilers go to great lengths to enforce correct use, even brashly criticising the programmers style when appropriate.

1.4.2.5 *Erroneous patterns*   Alternative patterns in a match case must share the same set of variable bindings. For example, although the following function makes sense to a human, the F# compilers complain about the patterns (a, 0) and (0, b) binding different sets of variables ($\{a\}$ and $\{b\}$, respectively):

```
> let product a b =
    match a, b with
    | a, 0 | 0, b -> 0
    | a, b -> a * b;;
Error: FS0018: The two sides of this 'or' pattern bind
different sets of variables
```

In this case, this function can be corrected by using the anonymous _ pattern as neither a nor b is used in the first case:

```
> let product a b =
    match a, b with
    | _, 0 | 0, _ -> 0
    | a, b -> a * b;;
val product : int -> int -> int
```

This actually conveys useful information about the code. Specifically, that the values matched by _ are not used in the corresponding expression.

F# uses type information to determine the possible values of expression being matched. If the patterns fail to cover all of the possible values of the input then, at compile-time, the compiler emits:

```
Warning: FS0025: Incomplete pattern match.
```

If a program containing such pattern matches is executed and no matching pattern is found at run-time then `MatchFailureException` is raised. Exceptions will be discussed in section 1.4.5.

For example, in the context of the built-in option type, the F# compiler will warn of a function matching only the `Some` type constructor and neglecting `None`:

```
> let extract = function
```

```
  | Some x -> x;;
Warning: FS0025: Incomplete pattern match.
The value 'None' will not be matched.
val extract : 'a option -> int
```

This `extract` function then works as expected when given that value `Some 3`:

```
> extract (Some 3);;
val it : int = 3
```

but causes `MatchFailureException` to be raised at run-time if a `None` value is given, as none of the patterns in the pattern match of the `extract` function match this value:

```
> extract None;;
Exception of type
'Microsoft.FSharp.MatchFailureException' was thrown.
```

As some approaches to pattern matching lead to more robust programs, some notions of good and bad programming styles arise in the context of pattern matching.

1.4.2.6 Good style The compiler cannot prove that any given pattern match covers all eventualities in the general case. Thus, some style guidelines may be productively adhered to when writing pattern matches, to aid the compiler in its proofs:

- Guarded patterns should be used only when necessary. In particular, in any given pattern matching, the last pattern should not be guarded.

- In the case of user-defined variant types, all eventualities should be covered explicitly (such as `[]` and `h::t` which, between them, match any list).

As proof generation cannot be automated in general, the F# compilers do not try to prove that a sequence of guarded patterns will match all possible inputs. Instead, the programmer is expected to adhere to a good programming style, making the breadth of the final match explicit by removing the guard. For example, the F# compilers do not prove that the following pattern match covers all possibilities:

```
> let sign = function
    | i when i < 0.0 -> -1
    | 0.0 -> 0
    | i when i > 0.0 -> 1;;
Warning: FS0025: Incomplete pattern match.
val sign : float -> int
```

In this case, the function should have been written without the guard on the last pattern:

```
> let sign = function
    | i when i < 0.0 -> -1
```

```
  | 0.0 -> 0
  | _ -> 1;;
val sign : float -> int
```

Also, the F# compilers will try to determine any patterns which can never be matched. If such a pattern is found, the compiler will emit a warning. For example, in this case the first match accounts for all possible input values and, therefore, the second match will never be used:

```
> let product a b =
    match a, b with
    | a, b -> a * b
    | _, 0.0 | 0.0, _ -> 0.0;;
Warning: FS0026: This rule will never be matched.
val product : int -> int -> int
```

When matching over the constructors of a type, all eventualities should be caught explicitly, i.e. the final pattern should not be made completely general. For example, in the context of a type which can represent different number representations:

```
> type number =
    | Integer of int
    | Real of float;;
type number = Integer of int | Real of float
```

A function to test for equality with zero could be written in the following, poor style:

```
> let bad_is_zero = function
    | Integer 0 | Real 0.0 -> true
    | _ -> false;;
val bad_is_zero : number -> bool
```

When applied to various values of type number, this function correctly acts a predicate to test for equality with zero:

```
> bad_is_zero (Integer (-1));;
val it : bool = false
> bad_is_zero (Integer 0);;
val it : bool = true
> bad_is_zero (Real 0.0);;
val it : bool = true
> bad_is_zero (Real 2.6);;
val it : bool = false
```

Although the bad_is_zero function works in this case, this formulation is fragile when the variant type is extended during later development of the program. Instead, the constructors of the variant type should be matched against explicitly, to ensure that later extensions to the variant type yield compile-time warnings for this function (which could then be fixed):

```
> let good_is_zero = function
    | Integer 0 | Real 0.0 -> true
    | Integer _ | Real _ -> false;;
val good_is_zero : number -> bool
```

The style used in the good_is_zero function is more robust. For example, if whilst developing our program, we were to supplement the definition of our number type with a new representation, say of the complex numbers $z = x + iy \in \mathbb{C}$:

```
> type number =
    | Integer of int
    | Real of float
    | Complex of float * float;;
type number =
    | Integer of int
    | Real of float
    | Complex of float * float
```

the bad_is_zero function, which is written in the poor style, would compile without warning despite being incorrect:

```
> let bad_is_zero = function
    | Integer 0 | Real 0.0 -> true
    | _ -> false;;
val bad_is_zero : number -> bool
```

Specifically, this function treats all values which are not zero-integers or zero-reals as being non-zero. Thus, zero-complex $z = 0 + 0i$ is incorrectly deemed to be non-zero:

```
> bad_is_zero (Complex (0.0, 0.0));;
val it : bool = false
```

In contrast, the good_is_zero function, which was written using the good style, would allow the compiler to spot that part of the number type was not being accounted for in the pattern match:

```
> let good_is_zero = function
    | Integer 0 | Real 0.0 -> true
    | Integer _ | Real _ -> false;;
Warning: FS0025: Incomplete pattern match.
val good_is_zero : number -> bool
```

The programmer could then supplement this function with a case for complex numbers:

```
> let good_is_zero = function
    | Integer 0 | Real 0.0 | Complex(0.0, 0.0) -> true
```

```
    | Integer _ | Real _ | Complex _ -> false;;
val good_is_zero : number -> bool
```

The resulting function would then provide the correct functionality:

```
> good_is_zero (Complex (0.0, 0.0));;
val it : bool = true
```

Clearly, the ability have such safety checks performed at compile-time can be very valuable during development. This is another important aspect of safety provided by the F# language, which results in considerably more robust programs.

Due to the ubiquity of pattern matching in F# programs, the number and structure of pattern matches can be non-trivial. In particular, patterns may be nested and may be performed in parallel.

1.4.2.7 *Parallel pattern matching*   Pattern matching is often applied to several different values in a single function. The most obvious way to pattern match over several values is to nest pattern matches. However, nested patterns are rather ugly and confusing.

For example, the following function tries to unbox three option types, returning None if any of the inputs is None:

```
> let unbox3 a b c =
    match a with
    | Some a ->
        match b with
        | Some b ->
            match c with
            | Some c -> Some(a, b, c)
            | None -> None
        | None -> None
    | _ -> None;;
val unbox3 :
  'a option -> 'b option -> 'c option ->
    ('a * 'b * 'c) option
```

Applying this function to three option values gives an option value in response:

```
> unbox3 (Some 1) (Some 2) (Some 3);;
val it : (int * int * int) option = Some (1, 2, 3)
```

Fortunately, parallel pattern matching can be used to perform the same task more concisely. This refers to the act of pattern matching over a tuple of values rather than nesting different pattern matches for each value.

For example, a function to unbox three option values simultaneously may be written more concisely using a parallel pattern match:

```
> let unbox3 a b c =
    match a, b, c with
```

```
    | Some a, Some b, Some c -> Some (a, b, c)
    | _ -> None;;
val unbox3 :
  'a option -> 'b option -> 'c option ->
    ('a * 'b * 'c) option
```

As a core feature of the F# language, pattern matching will be used extensively in the remainder of this book, particularly when dissecting data structures in chapter 3.

1.4.2.8 Active patterns ML-style pattern matching provides a simple and efficient way to dissect concrete data structures such as trees and, consequently, is ubiquitous in this family of programming languages. However, ML-style pattern matching has the disadvantage that it ties a function to a particular concrete data structure. A new feature in the F# programming language called *active patterns* is designed to alleviate this problem by allowing patterns to perform computations to dissect a concrete data structure and present it in a different form, known as a *view* of the underlying structure.

As a simple example, active patterns can be used to sanitize the strange total ordering function `compare` that F# inherited from OCaml by viewing the `int` result as the sum type that it really represents:

```
> let (|Less|Equal|Greater|) = function
    | c when c<0 -> Less
    | c when c>0 -> Greater
    | _ -> Equal;;
val (|Less|Equal|Greater|) :
  int -> Choice<unit, unit, unit>
```

Pattern matches over `int` values can now use the active patterns `Less`, `Equal` and `Greater`. Moreover, the pattern matcher is now aware that these three patterns form a complete set of alternatives.

A more useful example of active patterns is the dissecting of object oriented data structures carried over from the .NET world. The use of active patterns to simplify the dissection of XML trees is described in chapter 10.

1.4.3 Equality

The F# programming language includes a notion of structural equality that automatically traverses values of compound types such as tuples, records, variant types, lists and arrays as well as handling primitive types. The equality operator = calls the `Equals` method of the .NET object, allowing the equality operation to be overridden for specific types where applicable.

For example, the following checks that $3 - 1 = 2$:

```
> 3 - 1 = 2;;
val it : bool = true
```

The following tests the contents of two pairs for equality:

```
> (2, 3) = (3, 4);;
val it : bool = false
```

In some cases, the built-in structural equality is not the appropriate notion of equality. For example, the set data structure (described in detail in chapter 3) is represented internally as a balanced binary tree. However, some sets have degenerate representations, e.g. they may be balanced differently but the contents are the same. So structural equality is not the correct notion of equality for a set. Consequently, the Set module overrides the default Equals member to give the = operator an appropriate notion of set equality, where sets are compared by the elements they contain regardless of how they happen to be balanced.

Occasionally, the ability to test if two values refer to identical representations (e.g. the same memory location) may be useful. This is known as *reference* equality in the context of .NET and is provided by the built-in == operator.

For example, pairs defined in different places will reside in different memory locations. So, in the following example, the pair a is referentially equal to itself but a and b are logically but not referentially equal:

```
> let a = 1, 2;;
val a : int * int

> let b = 1, 2;;
val b : int * int

> a == a;;
val it : bool = true
stdin(26,1): warning: FS0062: This construct is for
compatibility with OCaml. The use of the physical
equality operator '==' is not recommended except in
cross-compiled code. You probably want to use generic
structural equality '='. Disable this warning using
--no-warn 632 or #nowarn "62"
```

This warning is designed for programmers used to languages where == denotes ordinary equality. Safe in the knowledge that == denotes referential equality in F#, we can disable this warning before using it:

```
> a = b;;
val it : bool = true

> #nowarn "62";;

> a == b;;
val it : bool = false
```

Referential equality may be considered a probabilistic alternative to logical equality. If two values are referentially equal then they must also be logically equal, otherwise they may or may not be logically equal. The notion of referential equality can be used to implement productive optimizations by avoiding unnecessary copying and is discussed in chapter 8.

1.4.4 Sequence expressions

The F# programming language provides an elegant syntax called *sequence expressions* for generating lists, arrays and Seq[5]. A contiguous sequence of integers can be specified using the syntax:

seq {*first* .. *last*}

For example, the the integers $i \in \{1 \ldots 5\}$ may be created using:

```
> seq {1 .. 5};;
val it : seq<int> = seq [1; 2; 3; ...]
```

Note that the generic sequence type is written using the syntax seq<'a> by default rather than 'a seq.

All comprehension syntaxes can be used with different brackets to generate lists and arrays. For example:

```
> [1 .. 5];;
val it : int list = [1; 2; 3; 4; 5]
> [|1 .. 5|];;
val it : int array = [|1; 2; 3; 4; 5|]
```

Non-contiguous sequences can also be created by specifying a step size using the syntax:

seq {*first* .. *step* .. *last*}

For example, the integers $[0 \ldots 9]$ in steps of 3 may be created using:

```
> seq {0 .. 3 .. 9};;
val it : seq<int> = seq [0; 3; 6; ...]
```

Lists, arrays and sequences can be filtered into a sequence using the syntax:

seq {for *pattern* in *container* ->
    expr}

For example, the squares of the Some values in an option list may be filtered out using:

```
> seq {for Some i in [Some 1; Some 3; None ; Some 2] ->
        i * i};;
val it : seq<int> = seq [1; 9; 4]
```

The pattern used for filtering can be guarded using the syntax:

seq {for *pattern* in *container* when *guard* ->
    expr}

For example, extracting only results for which $i < 3$ in the previous example:

---

[5]Seq is discussed in detail in section 3.8.

```
> let xs = [Some 1; Some 3; None; Some 2] in
  seq {for Some i in xs when i < 3 ->
        i * i};;
val it : seq<int> = seq [1; 4]
```

These examples all generate a data structure called Seq. This data structure is discussed in detail in chapter 3.

Comprehensions may also be nested to produce a flat data structure. For example, nesting loops over x and y coordinates is an easy way to obtain a sequence of grid coordinates:

```
> [ for x in 1 .. 3
      for y in 1 .. 3 ->
        x, y ];;
val it : (int * int) list =
  [1, 1; 1, 2; 1, 3; 2, 1; 2, 2; 2, 3; 3, 1; 3, 2; 3, 3]
```

Sequence expressions have a wide variety of uses, from random number generators to file IO.

1.4.5 Exceptions

In many programming languages, program execution can be interrupted by the raising[6] of an exception. This is a useful facility, typically used to handle problems such as failing to open a file or an unexpected flow of execution (e.g. due to a program being given invalid input).

Like a variant constructor in F#, the name of an exception must begin with a capital letter and an exception may or may not carry an associated value. Before an exception can be used, it must declared. An exception which does not carry associated data may be declared as:

exception *Name*

An exception which carries associated data of type *type* may be declared:

exception *Name* of *type*

Exceptions are raised using the built-in raise function. For example, the following raises a built-in exception called Failure which carries a string:

raise (Failure "My problem")

F# exceptions may be caught using the syntax:

try
  expr
with
| *pattern*₁ -> *expr*₁

---

[6]Sometimes known as *throwing* an exception, e.g. in the context of the C++ language.

| $pattern_2$ -> $expr_2$
| $pattern_3$ -> $expr_3$
| . . .
| $pattern_n$ -> $expr_n$

where *expr* is evaluated and its result returned if no exception was raised. If an exception was raised then the exception is matched against the patterns and the value of the corresponding expression (if any) is returned instead.

For example, the following raises and catches the `Failure` exception and returns the string that was carried by the exception:

```
> try
    raise (Failure "My problem")
  with
  | Failure s ->
      s;;
val it : string = "My problem"
```

Note that, unlike other pattern matching constructs, patterns matching over exceptions need not account for all eventualities — any uncaught exceptions simply continue to propagate.

For example, an exception called `ZeroLength` that does not carry associated data may be declared with:

```
> exception ZeroLength;;
exception ZeroLength
```

A function to normalize a 2D vector $\mathbf{r} = (x, y)$ to create a unit-length 2D vector:

$$\hat{\mathbf{r}} = \frac{\mathbf{r}}{|\mathbf{r}|}$$

Catching the erroneous case of a zero-length vector, this may be written:

```
> let norm (x, y) =
    match sqrt(x * x + y * y) with
    | 0.0 -> raise ZeroLength
    | s -> x / s, y / s;;
val norm : float * float -> float * float
```

Applying the `norm` function to a non-zero-length vector produces the correct result to within numerical error (a subject discussed in chapter 4):

```
> norm (3.0, 4.0);;
val it : float * float = (0.6, 0.8)
```

Applying the `norm` function to the zero vector raises the `ZeroLength` exception:

```
> norm (0.0, 0.0);;
Exception of type 'FSI_0159+ZeroLength' was thrown.
```

A "safe" version of the norm function might catch this exception and return some reasonable result in the case of a zero-length vector:

```
> let safe_norm r =
    try
      norm r
    with
    | ZeroLength ->
        0.0, 0.0;;
val safe_norm : float * float -> float * float
```

Applying the safe_norm function to a non-zero-length vector causes the result of the expression norm r to be returned:

```
> safe_norm (3.0, 4.0);;
val it : float * float = (0.6, 0.8)
```

However, applying the safe_norm function to the zero vector causes the norm function to raise the ZeroLength exception which is then caught within the safe_norm function which then returns the zero vector:

```
> safe_norm (0.0, 0.0);;
val it : float * float = (0.0, 0.0)
```

The use of exceptions to handle unusual occurrences, such as in the safe_norm function, is one important application of exceptions. This functionality is exploited by many of the functions provided by the core F# library, such as those for handling files (discussed in chapter 5). The safe_norm function is a simple example of using exceptions that could have been written using an if expression. However, exceptions are much more useful in more complicated circumstances, where an exception might propagate through several functions before being caught.

Another important application is the use of exceptions to escape computations. The usefulness of this way of exploiting exceptions cannot be fully understood without first understanding data structures and algorithms and, therefore, this topic will be discussed in much more detail in chapter 3 and again, in the context of performance, in chapter 8.

The Exit, Invalid_argument and Failure exceptions are built-in, as well as two functions to simplify the raising of these exceptions. Specifically, the invalid_arg and failwith functions raise the Invalid_argument and Failure exceptions, respectively, using the given string.

F# also provides a try ... finally ... construct that executes a final expression whether or not an exception is raised. This can be used to ensure that state changes are correctly undone even under exceptional circumstances.

1.5 IMPERATIVE PROGRAMMING

Just like conventional programming languages, F# supports mutable variables and side effects: imperative programming. Record fields can be marked as mutable, in which case their value may be changed. For example, the type of a mutable, two-dimensional vector called vec2 may be defined as:

```
> type vec2 = { mutable x: float; mutable y: float };;
type vec2 = { mutable x : float; mutable y : float; }
```

A value r of this type may be defined:

```
> let r = { x = 1.0; y = 2.0 };;
val r : vec2
```

The x-coordinate of the vector r may be altered in-place using an imperative style:

```
> r.x <- 3.0;;
val it : unit = ()
```

The side-effect of this expression has mutated the value of the variable r, the x-coordinate of which is now 3 instead of 1:

```
> r;;
val it : vec = {x = 3.0; y = 2.0}
```

A record with a single, mutable field can often be useful. This data structure, called a *reference*, is already provided by the type ref. For example, the following defines a variable named a that is a reference to the integer 2:

```
> let a = ref 2;;
val a : int ref = {contents = 2}
```

The type of a is then int ref. The value referred to by a may be obtained using !a:

```
> !a;;
val it : int = 2
```

The value of a may be set using :=:

```
> a := 3;;
val it : unit = ()
> !a;;
val it : int = 3
```

In the case of references to integers, two additional functions are provided, incr and decr, which increment and decrement references to integers, respectively:

```
> incr a;;
val it : unit = ()
> !a;;
val a : int = 4
```

In addition to mutable data structures, the F# language provides looping constructs for imperative programming. The while loop executes its body repeatedly while the condition is `true`, returning the value of type `unit` upon completion. For example, this `while` loop repeatedly decrements the mutable variable x, until it reaches zero:

```
> let x = ref 5;;
val x : int ref
> while !x > 0 do
    decr x;;
val it : unit = ()
> !x;;
val it : int = 0
```

The `for` loop introduces a new loop variable explicitly, giving the initial and final values of the loop variable. For example, this `for` loop runs a loop variable called i from one to five, incrementing the mutable value x five times in total:

```
> for i = 1 to 5 do
    incr x;;
val it : unit = ()
> !x;;
val it : int = 5
```

Thus, `while` and `for` loops in F# are analogous to those found in most imperative languages.

1.6 FUNCTIONAL PROGRAMMING

Unlike the *imperative* programming languages C, C++, C#, Java and Fortran, F# is a *functional* programming language. Functional programming is a higher-level and mathematically more elegant approach to programming that is ideally suited to scientific computing. Indeed, most scientists do not realise that they naturally compose programs in a functional style even if they are using an imperative language. We shall now examine the various aspects of functional programming and their implications in more detail.

1.6.1 Immutability

In mathematics, once a variable is defined to have a particular value, it keeps that value indefinitely. Thus, variables in mathematics are *immutable*. Similarly, most variables in F# are immutable.

In practice, the ability to choose between imperative and functional styles when programming in F# is very productive. Many programming tasks are naturally suited to either an imperative or a functional style. For example, portions of a program dealing with user input, such as mouse movements and key-presses, are likely to benefit from an imperative style where the program maintains a state and user input may result in a change of state. In contrast, functions dealing with the manipulation of complex data structures, such as trees and graphs, are likely to benefit from being written in a functional style, using recursive functions and immutable data, as this greatly simplifies the task of writing such functions correctly. In both cases, functions can refer to themselves — *recursive* functions. However, recursive functions are pivotal in functional programming, where they are used to implement functionality equivalent to the `while` and `for` looping constructs we have just examined.

One of the simplest differences between conventional imperative languages and functional programming languages like F# is the ubiquitous use of immutable data structures in functional programming. Indeed, the F# standard library provides a wealth of efficiently-implemented immutable data structures. The use of immutable data structures has some subtle implications and important benefits.

When a function acts upon an immutable data structure to produce a similar immutable data structure there is no need to copy the parts of the input that are reused in the output because the input data structure can never be changed. This ability to refer back to old values is known as *referential transparency*. So functions that act over immutable data structures typically compose an output that refers back to parts of the input.

For example, creating a list b as an element prepended onto a list a does not alter a:

```
> let a = [1; 2; 3];;
val a : int list
> let b = 0::a;;
val b : int list
> a, b;;
val it : int list * int list = ([1; 2; 3], [0; 1; 2; 3])
```

Note that the original list a is still [1; 2; 3]. Imperative programming would either require that a is altered (losing its original value) or that a is copied. Essentially, the former is confusing and the latter is slow.

Immutable data structures are beneficial for two main reasons:

- Simplicity: Mathematical expressions can often be translated into efficient functional programs much more easily than into efficient imperative programs.

- Concurrent: Immutable data structures are inherently thread safe so they are ideal for parallel programming.

When a programmer is introduced to the concept of functional programming for the first time, the way to implement simple programming constructs such as loops does not appear obvious. If the loop variable cannot be changed because it is immutable then how can the loop proceed?

1.6.2 Recursion

Looping constructs can be converted into recursive constructs, such as recurrence relations. For example, the factorial function is typically considered to be a product with the special case $0! = 1$:

$$n! = \prod_{i=1}^{n} i$$

This may be translated into an imperative function that loops over $i \in \{1 \ldots n\}$, altering a mutable accumulator called `accu`:

```
> let factorial n =
    let accu = ref 1
    for i = 1 to n do
      accu := i * !accu
    !accu;;
val factorial : int -> int
```

For example, $5! = 120$:

```
> factorial 5;;
val it : int = 120
```

However, the factorial may be expressed equivalently as a recurrence relation:

$$0! = 1$$
$$n! = n \times (n-1)!$$

This may be translated into an recursive function that calls itself until the base case $0! = 1$ is reached:

```
> let rec factorial = function
    | 0 -> 1
    | n -> n * factorial (n - 1);;
val factorial : int -> int
> factorial 5;;
val it : int = 120
```

In this case, the functional style is significantly simpler than the imperative style. As we shall see in the remainder of this book, functional programming is often more concise and simpler than imperative programming. This is particularly true in the context of mathematical programs.

The remaining aspects of functional programming are concerned with passing functions to functions and returning functions from functions.

1.6.3   Curried functions

A curried function is a function that returns a function as its result. Curried functions are best introduced as a more flexible alternative to the conventional (non-curried) functions provided by imperative programming languages.

Effectively, imperative languages only allow functions to accept a single value (often a tuple) as an argument. For example, a raise-to-the-power function for integers would have to accept a single tuple as an argument which contained the two values used by the function:

```
> let rec ipow_1(x, n) =
    match n with
    | 0 -> 1.0
    | n -> x * ipow_1(x, n - 1);;
val ipow_1 : float * int -> float
```

But, as we have seen, F# also allows:

```
> let rec ipow_2 n x =
    match n with
    | 0 -> 1.0
    | n -> x * ipow_2 (n - 1) x;;
val ipow_2 : int -> float -> float
```

This latter approach is actually a powerful generalization of the former, only available in functional programming languages.

The difference between these two styles is subtle but important. In the latter case, the type can be understood to mean:

```
val ipow_2 : int -> (float -> float)
```

i.e. this ipow_2 function accepts an exponent n and returns a function that raises a float x to the power of n, i.e. this is a curried function.

The utility of curried functions lies in their ability to have their arguments *partially applied*.

In this case, the curried ipow_2 function can have the power n partially applied to obtain a more specialized function for raising a float to a particular power. For example, functions to square and cube a float may now be written very succinctly in terms of ipow_2:

```
> let square = ipow_2 2;;
val square : float -> float
> square 5.0;;
val it : float = 25.0
> let cube = ipow_2 3;;
val cube : float -> float
> cube 3.0;;
val it : float = 27.0
```

Thus, the use of currying has allowed an expression of the form:

```
fun x -> ipow_1(x, 2)
```

to be replaced with the more succinct alternative:

```
ipow_2 2
```

This technique actually scales very well to more complicated situations with several curried arguments being partially applied one after another. As we shall see in the next chapter, currying is particularly useful when used in combination with higher-order functions.

1.6.4 Higher-order functions

Conventional languages vehemently separate functions from data. In contrast, F# allows the seamless treatment of functions as data. Specifically, F# allows functions to be stored as values in data structures, passed as arguments to other functions and returned as the results of expressions, including the return-values of functions.

A *higher-order function* is a function that accepts another function as an argument. As we shall now demonstrate, this ability can be of direct relevance to scientific applications.

Many numerical algorithms are most obviously expressed as one function parameterized over another function. For example, consider a function called d that calculates a numerical approximation to the derivative of a given one-argument function. The function d accepts a function $f : \mathbb{R} \to \mathbb{R}$ and a value x and returns a function to compute an approximation to the derivative $\frac{df}{dx}$ given by:

$$ d[f](x) = \frac{f(x + \delta) - f(x - \delta)}{2\delta} \simeq \frac{df}{dx} $$

where $d : (\mathbb{R} \to \mathbb{R}) \to (\mathbb{R} \to \mathbb{R})$.

This is easily written in F# as the higher-order function d that accepts the function f as an argument[7]:

```
> let d (f : float -> float) x =
    let dx = sqrt epsilon_float
    (f (x + dx) - f (x - dx)) / (2.0 * dx);;
val d : (float -> float) -> float -> float
```

For example, consider the function $f(x) = x^3 - x - 1$:

```
> let f x = x ** 3.0 - x - 1.0;;
val f : float -> float
```

The higher-order function d can be used to approximate $\frac{df}{dx}\Big|_{x=2} = 11$:

```
> d f 2.0;;
val it : float = 11.0
```

More importantly, as d is a curried function, we can use d to create derivative functions. For example, the derivative $f'(x) = \frac{\partial f}{\partial x}$ can be obtained by partially applying the curried higher-order function d to f:

---

[7]The built-in value epsilon_float is the smallest floating-point number that, when added to 1, does not give 1. The square root of this value can be shown to give optimal properties when used in this way.

```
> let f' = d f;;
val f' : (float -> float)
```

The function f' can now be used to calculate a numerical approximation to the derivative of f for any x. For example, $f'(2) = 11$:

```
> f' 2.0;;
val it : float = 11.0
```

Higher-order functions are invaluable for representing many operators found in mathematics, science and engineering.

Now that the foundations of F# have been introduced, the next chapter describes how these building blocks can be structured into working programs.

CHAPTER 2

PROGRAM STRUCTURE

In this chapter, we introduce some programming paradigms designed to improve program structure. As the topics addressed by this chapter are vast, we shall provide only overviews and then references to literature containing more thorough descriptions and evaluations of the relative merits of the various approaches.

Structured programming is all about managing complexity. Modern computational approaches in all areas of science are intrinsically complicated. Consequently, the efficient structuring of programs is vitally important if this complexity is to be managed in order to produce robust, working programs.

Historically, many different approaches have been used to facilitate the structuring of programs. The simplest approach involves splitting the source code of the program between several different files, known as *compilation units*. A marginally more sophisticated approach involves the creation of *namespaces*, allowing variable and function names to be categorised into a hierarchy. More recently, an approach known as *object-oriented* (**OO**) programming has become widespread. As we shall see, the F# language supports all of these approaches as well as others. Consequently, F# programmers are encouraged to learn the relative advantages and disadvantages of each approach in order that they may make educated decisions regarding the design of new programs.

F# for Scientists. By Jon Harrop
Copyright © 2008 John Wiley & Sons, Inc.

Structured programming is not only important in the context of large, complicated programs. In the case of simple programs, understanding the concepts behind structured programming can be instrumental in making efficient use of well-designed libraries.

2.1   NESTING

Function and variable definitions can be structured hierarchically within an F# program, allowing some definitions to be globally visible, others to be defined separately in distinct portions of the hierarchy, others to be visible only within a single branch of the hierarchy and others to be refined, specialising them within the hierarchy.

Compared to writing a program as a flat list of function and variable definitions, structuring a program into a hierarchy of definitions allows the number of dependencies within the program to be managed as the size and complexity of the programs grows. This is achieved most simply by *nesting* definitions. Thus, nesting is the simplest approach to the structuring of programs[8].

For example, the ipow3 function defined in the previous chapter contains a nested definition of a function sqr:

```
> let ipow3 x =
    let sqr x = x * x
    x * sqr x;;
val ipow3 : int -> int
```

The nested function sqr is only "visible" within the remainder of the definition of ipow3 and cannot be used outside this function. This capability is particularly useful when part of a function is best factored out into an auxiliary function that has no meaning or use outside its parent function. Nesting an auxiliary function can improve clarity without polluting the external namespace, e.g. with a definition ipow3_aux.

Nesting can be more productively exploited when combined with the factoring of subexpressions, functions and higher-order functions.

2.2   FACTORING

The concept of program factoring is best introduced in relation to the conventional factoring of mathematical expressions. When creating a complicated mathematical derivation, the ability to factor subexpressions, typically by introducing a substitution, is a productive way to manage the incidental complexity of the problem.

---

[8]Remarkably, many other languages, including C and C++, do not allow nesting of function and variable definitions.

2.2.1 Factoring out common subexpressions

For example, the following function definition contains several duplicates of the subexpression $x - 1$:

$$f(x) = (x - 1 - (x - 1)(x - 1))^{x-1}$$

By factoring out a subexpression $a(x)$, this expression can be simplified:

$$f(a) = (a - a^2)^a$$

where $a(x) = x - 1$.

The factoring of subexpressions, such as $x - 1$, is the simplest form of factoring available to a programmer. The F# function equivalent to the original, unfactored expression is:

```
> let f x =
    (x - 1.0 - (x - 1.0) * (x - 1.0)) ** (x - 1.0);;
val f : float -> float

> f 5.0;;
val it : float = 20736.0
```

The F# function equivalent to the factored form is:

```
> let f x =
    let a = x - 1.0
    (a - a * a) ** a;;
val f : float -> float

> f 5.0;;
val it : float = 20736.0
```

By simplifying expressions, factoring is a means to manage the complexity of a program. The previous example only factors a subexpression but, in functional languages such as F#, it is possible to factor out higher-order functions.

2.2.2 Factoring out higher-order functions

As we have just seen, the ability to factor out common subexpressions can be useful. However, functions are first-class values in functional programming languages like F# and, consequently, it is also possible to factor out common sub-functions. In fact, this technique turns out to be an incredibly powerful approach to software engineering that is of particular importance in the context of scientific computing.

Consider the following functions that compute the sum and the product of a semi-inclusive range of integers $[x_0, x_1)$, respectively:

```
> let rec sum_range x0 x1 =
    if x0 = x1 then 0 else x0 + sum_range (x0 + 1) x1;;
val sum_range : int -> int -> int
```

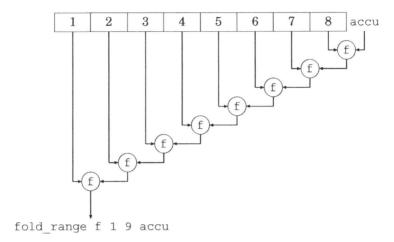

```
fold_range f 1 9 accu
```

Figure 2.1   The fold_range function can be used to accumulate the result of applying a function f to a contiguous sequence of integers, in this case the sequence $[1, 9)$.

```
> let rec product_range x0 x1 =
    if x0 = x1 then 1 else
      x0 * product_range (x0 + 1) x1;;
val product_range : int -> int -> int
```

For example, the product_range function may be used to compute 5! as the product of the integers $[1, 6)$:

```
> product_range 1 6;;
val it : int = 120
```

The sum_range and product_range functions are very similar. Specifically, they both apply a function (integer add + and multiply *, respectively) to 1 before recursively applying themselves to the smaller range $[x_0 + 1, x_1)$ until the range contains no integers, i.e. $x_0 = x_1$.

This commonality can be factored out as a higher-order function fold_range:

```
> let rec fold_range f x0 x1 accu =
    if x0 = x1 then accu else
      f x0 (fold_range f (x0 + 1) x1 accu);;
val fold_range :
(int -> 'a -> 'a) -> int -> int -> 'a -> 'a
```

The fold_range function accepts a function f, a range specified by two integers x0 and x1 and an initial value accu. Application of the fold_range function to mimic the sum_range or product_range functions begins with a base case in accu (0 or 1, respectively). If $x_0 = x_1$ then accu is returned as the result. Otherwise, the fold_range function applies f to both 1 and the result of a recursive call with the smaller range $[l + 1, u)$. This process, known as a *right* fold because f is applied to the rightmost integer and the accumulator first, is illustrated in figure 2.1.

The `sum_range` and `product_range` functions may then be expressed more simply in terms of the `fold_range` function by supplying a base case (0 or 1, respectively) and operator (addition or multiplication, respectively)[9]:

```
> let sum_range x0 x1 = fold_range ( + ) x0 x1 0;;
val sum_range : int -> int -> int
> let product_range x0 x1 = fold_range ( * ) x0 x1 1;;
val product_range : int -> int -> int
```

These functions work in exactly the same way as the originals:

```
> product_range 1 6;;
val it : int = 120
```

but their commonality has been factored out into the `fold_range` function. Note how succinct the definitions of the `sum_range` and `product_range` functions have become thanks to this factoring.

In addition to simplifying the definitions of the `sum_range` and `product_range` functions, the `fold_range` function may also be used in new function definitions. For example, the following higher-order function, given a length n and a function f, creates the list containing the n elements $(f(0), f(1), \ldots, f(n-1))$:

```
> let list_init n f =
    fold_range (fun h t -> f h :: t) 0 n [];;
val list_init : int -> (int -> 'a) -> 'a list
```

This `list_init` function uses the `fold_range` function with a λ-function, an accumulator containing the empty list `[]` and a range $[0, n)$. The λ-function prepends each `f h` onto the accumulator `t` to construct the result. For example, the `list_init` function can be used to create a list of squares:

```
> list_init 5 (fun i -> i*i);;
val it : int list = [0; 1; 4; 9; 16]
```

The functionality of this `list_init` function is already provided in the F# standard library in the form of the `List.init` function:

```
> List.init 5 (fun i -> i*i);;
val it : int list = [0; 1; 4; 9; 16]
```

In fact, this functionality of the `List.init` function can be obtained more clearly using sequence expressions and comprehensions (described in section 1.4.4):

```
> [ for i in 0 .. 4 ->
      i * i ];;
val it : int list = [0; 1; 4; 9; 16]
```

As we have seen, the nesting and factoring of functions and variables can be used to simplify programs and, therefore, to help manage intrinsic complexity. In

---

[9]The non-infix form of the + operator is written (+), i.e. (+) a b is equivalent to a + b. Note the spaces to avoid (*) from being interpreted as the start of a comment.

addition to these approaches to program structuring, the F# language also provides two different constructs designed to encapsulate program segments. We shall now examine the methodology and relative benefits of the forms of encapsulation offered by *modules* and *objects*.

2.3   MODULES

In the F# programming language, substantial sections of related code can be productively encapsulated into *modules*.

We have already encountered several modules. In particular, the predefined module:

```
Microsoft.FSharp.Core.Operators
```

encapsulates many function and type definitions initialized before a program is executed, such as operators on the built-in `int`, `float`, `bool` and `string` types. Also, the `Array` module encapsulates function and type definitions related to arrays, such as functions to create arrays (e.g. `Array.init`) and to count the number elements in an array (`Array.length`).

In general, modules are used to encapsulate related definitions, i.e. function, type, data, module and object definitions. In addition to simply allowing related definitions to be grouped together, the F# module system allows these groups to form hierarchies and allows interfaces between these groups to be defined, the correct use of which is then enforced by the compiler at compile-time. This can be used to add "safety nets" when programming, to improve program reliability. We shall begin by examining some of the modules available in the core F# library before describing how a program can be structured using modules, stating the syntactic constructs required to create modules and, finally, developing a new module.

By convention, if a module focuses on a single type definition (such as the definition of a set in the `Set` module) then that type is named `t`, e.g. `Set.t`.

In the interests of clarity and correctness, modules provide a well-defined interface known as a *signature*. The signature of a module declares the contents of the module which are accessible to code outside the module. The code implementing a module is defined in the module *structure*.

The concepts of module signatures and structures are best introduced by example. Consider a module called `FloatRange` that encapsulates code for handling ranges on the real number line, represented by a pair of `float` values. This module will include:

- A type `t` to represent a range.

- A function `make` to construct a range from a given `float` pair.

- A function `mem` to test a number for membership in a range.

- Set theoretic operations union and intersection.

A FloatRange module adhering to a given signature may be defined as:

```
> module FloatRange =
    type t = { x0: float; x1: float }
    let make x0 x1 =
      if x1 >= x0 then { x0 = x0; x1 = x1 } else
      invalid_arg "FloatRange.make"
    let mem x r =
      r.x0 <= x && x < r.x1
    let order a b =
      if a.x0 < b.x1 then b, a else a, b
    let union a b =
      let a, b = order a b
      if a.x1 < b.x0 then [a; b] else
      [make (min a.x0 b.x0) (max a.x1 b.x1)]
    let inter a b =
      let a, b = order a b
      if a.x1 < b.x0 then [] else
      [make (max a.x0 b.x0) (min a.x1 b.x1)]];;
module FloatRange : begin
  type t = {x0: float;
            x1: float;}
  val make : float -> float -> t
  val mem : float -> t -> bool
  val order : t -> t -> t * t
  val union : t -> t -> t list
  val inter : t -> t -> (float * float) list
end
```

This FloatRange module encapsulated definitions relating to ranges of real-valued numbers and may be used to create and perform operations upon values representing such ranges. For example, a pair of ranges may be created:

```
> let a, b =
    FloatRange.make 1.0 3.0, FloatRange.make 2.0 5.0;;
val a : FloatRange.t
val b : FloatRange.t
```

The union and intersection of these ranges may then be calculated using the union and inter functions provided by the FloatRange module. For example, $[1,3) \cup [2,5) = [1,5)$:

```
> FloatRange.union a b;;
val it : FloatRange.t list = [{x0 = 1.0; x1 = 5.0}]
```

and $[1,3) \cap [2,5) = [2,3)$:

```
> FloatRange.inter a b;;
val it : FloatRange.t list = [{x0 = 2.0; x1 = 3.0}]
```

The F# standard library provides many useful modules, such as the modules used to manipulate data structures described in chapter 3. In fact, the same syntax can be used to access objects, such as the *.NET* standard library.

Modules can be extended by simply creating a new module with the same name. As the two namespaces are indistinguishable, F# will look in all definitions of a module with a given name starting with the most recent definition.

2.4 OBJECTS

The .NET platform is fundamentally an object oriented programming environment. Related definitions are encapsulated in classes of objects under .NET. The F# programming language actually compiles all of the constructs we have seen (e.g. records, variants and modules) into .NET objects. This can be leveraged by augmenting these F# definitions with object-related definitions. Object oriented programming is particularly important when interoperating with .NET libraries, which will be discussed in chapter 9.

2.4.1 Augmentations

Properties and members are written using the syntax:

member *object*. *member args* = ...

Record and variant types may be augmented with member functions using the `type ... with ...` syntax. For example, a record representing a 2D vector may be supplemented with a member to compute the vector's length:

```
> type vec2 = { x: float; y: float } with
    member r.Length = sqrt(r.x * r.x + r.y * r.y);;
```

Note that the method name `r.Length` binds the current object to `r` in the body of the method definition.

The length of a value of type `vec2` may then be computed using the object oriented syntax:

```
> {x = 3.0; y = 4.0}.Length;;
val it : float = 5.0
```

The member `Length`, which accepts no arguments, is called a *property*.

2.4.1.1 Getters and Setters Properties can provide both `get` and `set` functions. This capability can be used to good effect in this case by allowing the length of a vector to be set:

```
> type vec2 = {mutable x: float; mutable y: float} with
```

```
member r.Length
  with get () =
    sqrt(r.x * r.x + r.y * r.y)
  and set len =
    let s = len / r.Length
    r.x <- s * r.x;
    r.y <- s * r.y;;
```

Note that the fields of the record have been marked mutable so that they can be set.

For example, setting the length of a vector to one scales the vector to unit length:

```
> let r = {x = 3.0; y = 4.0};;
val r : vec2
> r.Length <- 1.0;;
val it : unit = ()
> r;;
val it : vec2 = {x=0.6; y=0.8}
```

As this example has shown, significant functionality can be hidden or *abstracted away* using contructs like getters and setters. This ability to hide complexity is vital when developing complicated programs. The form of abstraction offered by getters and setters is most useful in the context of GUI programming, where it can be used to provide an editable interface to a GUI that hides the internal representation and communication required to reflect changes in a GUI.

2.4.1.2 *Indexing*   Classes, records and variants can also provide an Item property to allow values to be indexed. For example, we may wish to index the vec2 type by an integer 0 or 1:

```
> type vec2 = {mutable x: float; mutable y: float} with
    member r.Item
      with get(d) =
        match d with
        | 0 -> r.x
        | 1 -> r.y
        | _ -> invalid_arg "vec2.get"
      and set d v =
        match d with
        | 0 -> r.x <- v
        | 1 -> r.y <- v
        | _ -> invalid_arg "vec2.set";;
```

For example, setting the *x*-coordinate of **r** to 3:

```
> let r = {x=4.0; y=4.0};;
val r : vec2
```

```
> r.[0] <- 3.0;;
val it : unit = ()
> r;;
val it : vec2 = {x=3.0; y=4.0}
```

The vec2 type can now act as an ordinary container, like an array, but the dimensionality is constrained to be 2. Moreover, this constraint is checked by F#'s static type system. If we had a similar vec3 type then vectors of different dimensionalities will be considered incompatible by the F# type system and any errors confusing vectors of the two different types will be caught at compile time.

Leveraging the static type system in this way helps to catch bugs and improve program reliability and reduce development time. There are many ways that F#'s type system can be exploited and some more interesting methods will be described later in this book.

2.4.1.3 *Operator Overloading*

In the interests of brevity and clarity, many operators are overloaded in F#. For example, the + and * operators can be used to add numbers of several different types, including int and float. These overloaded operators can be extended to work on user defined types by augmenting the type with static member functions.

Static member functions are member functions that are not associated with any particular object and are defined using the syntax:

```
static member member args = ...
```

For example, our vec2 type may be supplemented with a definition of a static member to provide an overload of the built-in + operator for this type:

```
> type vec2 = { x: float; y: float } with
    static member ( + ) (a, b) =
      {x = a.x + b.x; y = a.y + b.y};;
```

The + operator can then be used to add vectors of the type vec2:

```
> {x = 2.0; y = 3.0} + {x = 3.0; y = 4.0};;
val it : vec2 = {x = 5.0; y = 7.0}
```

Many of the built-in operators can be overloaded. See the F# manual for specific details.

2.4.2 Classes

In conventional object-oriented programming languages, the vec2 type would have been defined as an ordinary class. There are two different ways to define ordinary classes in F#.

2.4.2.1 *Explicit constructors*

The following definition of the vec2 class uses the conventional style with private variables x_ and y_, members x and y that expose their values publically, two overloaded constructors and a Length property that uses the public members rather than the private variables:

```
> type vec2 =
    val private x_  : float
    val private y_  : float
    new(x)  =
      {x_ = x; y_ = 0.0}
    new(x, y)  =
      {x_ = x; y_ = y}
    member r.x = r.x_
    member r.y = r.y_
    member r.Length =
      sqrt(r.x * r.x + r.y * r.y);;
```

Note the verbosity of this approach compared to the augmented record definition (section 2.4.1).

This two overloaded constructors are defined using the new keyword. The first constructs a vec2 assuming the default $y = 0$. The second constructs a vec2 from both x and y coordinates.

The public members x and y facilitate abstraction and by using these, rather than the internal private variables x_ and y_ directly, the Length property will continue to work if the internal representation is changed. For example, the internal variables x_ and y_ might be altered to represent the vector is polar coordinates, in which case the public members x and y will be updated to continue to present the vector in cartesian coordinates and the Length property will continue to function correctly. This might be useful if performance is limited by operations that are faster in the polar representation.

In F#, the new keyword is not required when constructing an object. For example, the following creates a vec2 using the second constructor and calculates its length using the Length property:

```
> vec2(3.0, 4.0).Length;;
val it : float = 5.0
```

The F# programming language also provides a more concise and functional approach to object definitions when there is a single implicit constructor.

2.4.2.2 Implicit Constructor   Many classes present only one constructor that is used to create all objects of the class. In F#, a shorthand notation exists that allows an implicit constructor to be defined from a more idiomatic F# definition of the class using let bindings. The vec2 example may be written with the arguments of the constructor in the first line of the type definition:

```
> type vec2(x : float, y : float) =
    member r.x = x
```

```
member r.y = y
member r.Length =
  sqrt(r.x * r.x + r.y * r.y);;
```

This shorter definition works in exactly the same way as the previous one except that it provides only one constructor:

```
> vec2(3.0, 4.0).Length;;
val it : float = 5.0
```

This object-oriented style of programming is useful when dealing with the interface between F# and other .NET languages. Object oriented programming will be used later in this book for exactly this purpose, particularly when dealing with DirectX and Windows Forms GUI programming in chapter 7 and .NET interoperability in chapters 9 and 11.

2.4.2.3 Run-time type testing

The F# programming language inherits the use of run-time type information from the underlying .NET platform. This allows interfaces to be dynamically typed and allows the value of a type to be determined at run time. Testing of run-time types is virtually non-existent in self-contained idiomatic F# code but this feature can be very important in the context of interfacing to other .NET languages and libraries. In particular, run-time type testing is used in F# programs to catch general .NET exceptions.

In F#, run-time type tests appear as a special pattern matching construct : ? written in the form:

```
match object with
| :? class -> ...
```

In this case, it is often useful to extract the matched value using a named subpattern (this technique was described in section 1.4.2.2):

```
match object with
| :? class as x -> f x
```

The ability to match values of a particular type is important when handling exceptions generated by other .NET programs or libraries. This technique is used in chapter 7 to handle exceptions generated by the DirectX library.

2.4.2.4 Boxing

Excessive use of objects is a serious impediment to good performance. Consequently, the .NET platform provides several *primitive* types that are not represented by objects, such as `int` and `float`. In some cases, it can be necessary to force a value of a primitive type into an object oriented representation where its class is derived from the univeral `obj` class. In F#, this functionality is provided by the `box` function. For example, boxing an `int`:

```
> box 12;;
val it : obj = 12
```

The ability to represent any value as a type derived from `obj` is essentially dynamic typing. The use of dynamic typing has many disadvantages, most notably performance degredation and unreliability due to the lack of static checking by the compiler. Consequently, use of dynamic typing should be minimized, limited only to cases where existing interfaces require it to be used. This functionality is discussed later in the context of the DirectX bindings (chapter 7) and the .NET metaprogramming facilities (chapter 9).

2.5 FUNCTIONAL DESIGN PATTERNS

Like all programming languages, programs written in F# often contain idiomatic approaches to solving certain types of problem or expressing particular functionality. Many books have been written on design patterns in imperative procedural or object oriented languages but there is little information about design patterns in functional programming and nothing specifically on F#.

This section describes some common styles adopted by F# programmers that will be used in the remainder of this book. The clarity and elegance of these approaches makes them worth learning in their own right.

2.5.1 Combinators

The ability to handle functions as values opens up a whole new way of programming. When making extensive use of functional values, the ability to compose these values in different ways can be very useful.

Higher-order functions and currying are often combined to maximize the utility of function composition. Higher-order functions that work by composing applications of other functions (including their argument functions) are sometimes referred to as *combinators* [13]. Such functions can be very useful when designed appropriately.

The practical benefits of combinators are most easily elucidated by example. In fact, we have already seen some combinators such as the derivative function `d` (from section 1.6.4).

The F# language provides three useful combinators that are written as operators:

- `f << g` composes the functions f and g (written $f \circ g$ in mathematics) such that `(f << g) x` is equivalent to `f(g x)`.

- `f >> g` is a reverse function composition, such that `(f >> g) x` is equivalent to `g(f x)`.

- `x |> f` is function application written in reverse, equivalent to `f x`.

These combinators will be used extensively in the remainder of this book, particularly the `|>` combinator.

A surprising amount can be accomplished using combinators. A simple combinator that applies a function to its argument twice may be written:

```
> let apply_2 f =
    f << f;;
val apply_2 : ('a -> 'a) -> ('a -> 'a)
```

For example, doubling the number five twice is equivalent to multiplication by four:

```
> apply_2 (( * ) 2) 5;;
val it : int = 20
```

A combinator that applies a function four times may be written elegantly in terms of the existing `apply_2` combinator:

```
> let apply_4 f =
    apply_2 apply_2 f;;
val apply_4 : ('a -> 'a) -> ('a -> 'a)
```

As a generalization of this, a combinator can be used to *nest* many applications of the same function:

```
> let rec nest n f x =
    match n with
    | 0 -> x
    | n -> nest (n - 1) f (f x);;
val nest : int -> ('a -> 'a) -> 'a -> 'a
```

For example, `nest 3 f x` gives `f(f(f x))`:

```
> nest 3 (( * ) 2) 1;;
val it : int = 8
```

Recursion itself can be factored out into a combinator by rewriting the target function as a higher-order function that accepts the function that it is to call, rather than as a recursive function. The following y combinator can be used to reconstruct a one-argument recursive function:

```
> let rec y f x =
    f (y f) x;;
val y : (('a -> 'b) -> 'a -> 'b) -> 'a -> 'b
```

Note that the argument f to the y combinator is itself a combinator.

For example, the factorial function may be written in a non-recursive form, accepting another `factorial` function to call recursively as its first argument:

```
> let factorial factorial = function
    | 0 -> 1
    | n -> n * factorial(n - 1);;
val factorial : (int -> int) -> int -> int
```

This process of rewriting a recursive function as a combinator to avoid explicit recursion is known as "untying the recursive knot" and makes it possible to stitch functions together, injecting new functionality between function calls. For example, this will be used to implement recursive memoization in section 6.1.4.2.

Applying the y combinator yields the usual factorial function:

```
> y factorial 5;;
val it : int = 120
```

Composing functions using combinators makes it easy to inject new code in between function calls. For example, a function to print the argument can be injected before each invocation of the `factorial` function:

```
> y (factorial >> fun f n -> printf "%d\n" n; f n) 5;;
5
4
3
2
1
0
val it : int = 120
```

This style of programming is useful is many circumstances but is particularly importance in the context of asynchronous programming, where entire computations are formed by composing fragments of computation together using exactly this technique. This is so useful that the F# language recently introduced a customized syntax for asynchronous programming known as *asynchronous workflows*.

The task of measuring the time taken to apply a function to its argument can be productively factored into a `time` combinator. The .NET `Stopwatch` class can be used to measure elapsed real time with roughly millisecond (0.001s) accuracy. The following `time` combinator accepts a function f and its argument x and times how long it takes to apply f to x, printing the time taken and returning the result f x:

```
> let time f x =
    let timer = new System.Diagnostics.Stopwatch()
    timer.Start()
    try f x finally
    printf "Took %dms\n" timer.ElapsedMilliseconds;;
val time : ('a -> 'b) -> 'a -> 'b
```

Note the use of `try .. finally` to ensure that the time taken is printed even if the function application f x raises an exception.

For example, the time taken to construct a 10^6-element list:

```
> time (fun () -> [1 .. 1000000]) ();;
Took 446ms
```

Computations can also be timed from F# interactive sessions by executing the `#time` statement:

```
> #time;;
--> Timing now on
```

```
> [1 .. 1000000];;
Real: 00:00:00.619, CPU: 00:00:00.640, GC gen0: 6,
gen1: 2, gen2: 1
> #time;;
--> Timing now off
```

This has the advantage that it prints both elapsed real time and CPU time (which will be discussed in section 8) as well as the number of different generations of garbage collection that were invoked (quantifying how much the computation stressed the garbage collector). However, our `time` combinator has the advantage that it can be placed inside computations uninvasively in order to time certain parts of a computation.

Our `time` combinator will be used extensively throughout this book to compare the performance of different functions. In particular, chapter 8 elaborates on the `time` combinator in the context of performance measurement, profiling and optimization.

2.5.2   Maps and folds

Section 2.2 introduced the concept of a fold. This is a higher-order function that applies a given function argument to a set of elements, accumulating a result. In the previous example, the fold acted upon an implicit range of integers but folds are most often associated with container types (such as lists and arrays), a subject covered in detail in chapter 3.

Folds over data structures are often categorized into left and right folds, referring to starting with the left-most (first) or starting with the right-most (last) element, respectively.

The addition operator can be folded over a container in order to sum the int elements of a container `xs`. With a left fold this is written in the form:

```
fold_left ( + ) 0 xs
```

With a right fold this is written in the form:

```
fold_right ( + ) xs 0
```

Note the position of the initial accumulator 0, on the left for a left-fold and on the right for a right-fold.

For example, if `xs` is the sequence $(1, 2, 3)$ then the left-fold computes $((0 + 1) + 2) + 3$ whereas a right-fold computes $1 + (2 + (3 + 0))$.

Maps are a similar concept. These higher-order functions apply a given function argument f to a set of elements x_i to create a new set of elements $y_i = f(x_i)$. Container types (e.g. lists and arrays) typically provide a `map` function.

Maps and folds are generic algorithms used in a wide variety of applications. The data structures in the F# standard library provide maps and folds where appropriate, as we shall see in chapter 3.

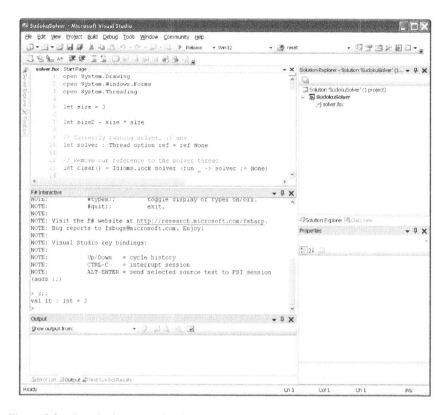

Figure 2.2   Developing an application written entirely in F# using Microsoft Visual Studio 2005.

2.6   F# DEVELOPMENT

The simplest way to develop software in F# is to use Microsoft's own Visual Studio and then install Microsoft's F# distribution. Visual Studio is an integrated development environment that makes it easy to develop *.NET* applications in various languages including F#. Moreover, Visual Studio allows you to build applications that are written in several different programming languages. Figure 2.2 shows one of the applications from The F#.NET Journal being developed using Visual Studio 2005.

In Visual Studio terminology, an application or program is a *solution* and a solution may be composed of several *projects*. Projects may be written in different languages and refer to each other. Both individual projects and complete solutions can be *built* to create executables that can then be run.

```
type vec2 = { mutable x: float; mutable y: float } with
  member r.Item
    with get = function
       | 0 -> r.x
       | 1 -> r.y
       | _ -> invalid_arg "vec2.get"
    and set i v = match i with
       | 0 -> r.x <- v
       | 1 -> r.y <- v
       | _ -> invalid_arg "vec2.set"
end;;

let r = {x=4.; y=4.};;
     val r : vec2
r.[0] <- 3.;;
```

Figure 2.3   Visual Studio provides graphical throwback of the type information inferred by the F# compiler: hovering the mouse over the definition of a variable r in the source code brings up a tooltip giving the inferred type of r.

2.6.1   Creating an F# project

To create a new project in Visual Studio click File ▷ New ▷ Project. From the New Project window, select F# Projects from the Project type tree, give the project a name and click OK.

To add an F# source code file to the new project, right-click on the project's name in the Solution Explorer pane and select Add ▷ New Item. From the Add New Item window, select F# Source File, give the file a name and click OK. The new F# source code file can then be edited.

An important advantage of Visual Studio is the graphical throwback of inferred type information. For example, figure 2.3 shows the inferred type vec2 of the variable r in some F# source code. As type information is not explicit in F#, this functionality is extremely useful when developing F# programs.

Once written, F# source code can be executed in two different ways: a solution or project may be built to create an executable or the code may be entered into a running F# interactive mode.

2.6.2   Building executables

A Visual Studio solution or project may be built to create an executable. To build an executable press CTRL+SHIFT+B. More often, solutions are built and executed immediately. This can be done in one step by pressing F5 to build and run a solution in *debug* mode or CTRL+F5 to build and run a solution in *release* mode. As the

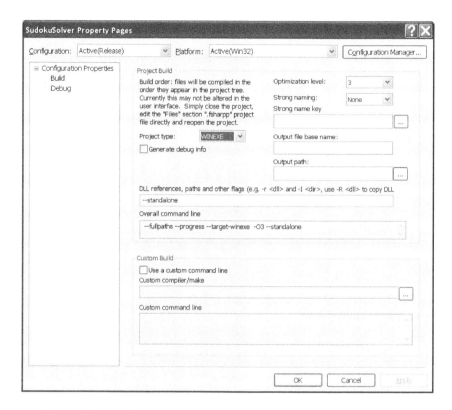

Figure 2.4 A project's properties page allows the compiler to be controlled.

names imply, debug mode can provide more information when a program goes wrong as opposed to release mode which executes more quickly.

Various options can be adjusted in the Project Properties window (see figure 2.4), to control the way an executable is built:

- If an executable is to be distributed to third parties (e.g. when writing a commercial application), the project should be compiled with the option --standalone in order to compile in the F# libraries that are not included in *.NET* itself.

- Graphical programs do not want the console window that appears by default. This can be removed either by changing the project type from EXE to WINEXE or by compiling with the option --target-winexe.

- Libraries used by a project may be described on the compile line using the syntax -r *name*.dll or in the source code using the #r pragma.

These options are set in the Project Properties window.

Although a detailed tutorial on Visual Studio is beyond the scope of this book it is worth mentioning the subject of debugging.

2.6.3 Debugging

The static typing provided by the F# language goes a long way in eliminating errors in programs before they are run. However, there are still cases when a program will fail during its execution. In such cases, a detailed report of the problem can be essential in debugging the application.

Visual Studio provides a great deal of debugging support to help the programmer. In the case of F#, an application that fails (typically due to the raising of an exception that is never caught and handled) gives immediate throwback on where the exception came from and what it contained. Breakpoints can be set the in source code, to stop execution when a particular point is reached. The values of variables may then be inspected.

One of the most useful capabilities of the Visual Studio debugger is the ability to pause a program when an exception is thrown even if it is subsequently caught. This functionality is enabled through the Debug ▷ Exceptions... dialog box.

In addition to building executable applications, F# provides a powerful interactive mode that allows code to be executed without being built beforehand.

2.6.4 Interactive mode

In addition to conventional project building, F# provides an interactive way to execute code, called the F# interactive mode. This mode is useful for trying examples when first learning the F# programming language and for testing snippets of code from programs during development.

Scientists and engineers typically use two separate kinds of programming language:

1. Fast compiled languages like C and Fortran.

2. Expressive high-level interactive languages with integrated visualization like Matlab and Mathematica.

F# is the first language to combine the best of both worlds with the F# interactive mode.

Before the F# interactive mode can be used within Visual Studio, it must be selected in the Add-in Manager (available from the menu Tools ▷ Add-in Manager, see figure 2.5).

The F# interactive mode can be started by pressing ALT+ENTER in the editor window. Lines of code from the editor can be evaluated in an interactive mode by pressing ALT+'. Selected blocks of code can be evaluated in an interactive mode by pressing ALT+ENTER. When typing directly into an F# interactive session, evaluation is forced by ending the last line with ;;.

2.6.4.1 *Double semicolons*   The examples given so far in this book have been written in the style of the F# interactive mode and, in particular, have ended with ;; to force evaluation. These double semicolons are not necessary in compiled code (i.e. in .fs and .fsx files).

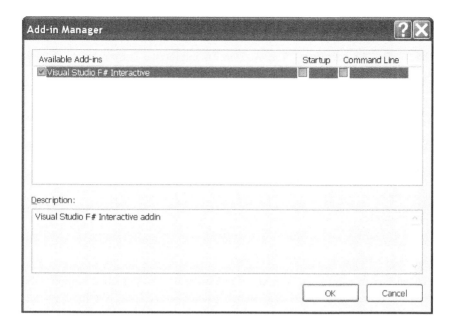

Figure 2.5   The Add-in Manager is used to provide the F# interactive mode.

For example, the two function definitions can be interpreted simultaneously in an interactive session, with only a single ; ; at the end:

```
> let f1 x =
    if x < 3 then print_endline "Less than three"
  let f2 x =
    if x < 3 then "Less" else "Greater";;
val f1 : int -> unit
val f2 : int -> string
```

When compiled, programs can be written entirely without ; ; .

2.6.4.2 Loading F# modules   Module definitions from compilation units (the .fs and .fsx files) can be loaded into a running F# interactive session using the #use and #load directives.

The #use directive interprets the contents of a compilation unit directly, defining everything from a module directly into the running interactive session. For example, assuming a file "foo.fs" contains a definition for x, invoking the #use directive:

```
> #use "foo.fs";;
```

causes the definition of x to be available from the interactive session.

The #load directive interprets the contents of a compilation unit as a module. For example:

```
> #load "foo.fs";;
```

makes the definition of x available as Foo.x.

2.6.4.3 *Pretty printing*

The way an F# interactive session prints values of different types can be altered by registering new functions to convert values into strings with the AddPrinter member function of the fsi object.

The following variant type might be used to represent lengths in different units, starting with one type constructor to represent lengths in meters:

```
> type length = Meters of float;;
```

Values of this type may be pretty printed with:

```
> fsi.AddPrinter(fun (Meters x) -> sprintf "%fm" x);;
```

For example:

```
> Meters 5.4;;
val it : length = 5.400000m
```

The ability to control the printing of values by an F# interactive session will be used in the remainder of this book. In particular, chapter 7 will use this approach is to spawn windows visualizing values that represent graphics.

2.6.5   C# interoperability

One of the great benefits of the common language run-time that underpins .NET is the ability to interoperate with dynamically linked libraries (**DLL**s) written in other .NET languages. In particular, this opens up the wealth of tools available for mainstream languages like C# including the GUI Windows Forms designers integrated into Visual Studio and the automatic compilation of web services into C# functions. A specific example of this will be discussed in chapter 10 in the context of web services.

An F# program that requires a C# library is easily created using Visual Studio:

1. Create a C# project inside a new Solution directory (see figure 2.6).

2. Create an F# project in the same solution (see figure 2.7).

3. Set the startup project of the solution to the F# program rather than the DLL (see figure 2.8) by right clicking on the solution in the "Solution Explorer" window and selecting Properties from the menu.

4. Build the C# DLL (right click on it in Solution Explorer and select "Rebuild") and reference it from the F# program.

The *name.dll* DLL can be referenced from the source code of an F# program with:

#r "*name.dll*"

In this case, the directory of the DLL must be added to the search path before the DLL is referenced.

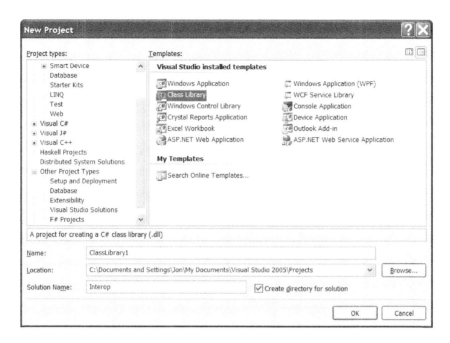

Figure 2.6   Creating a new C# class library project called `ClassLibrary1` inside a new solution called `Interop`.

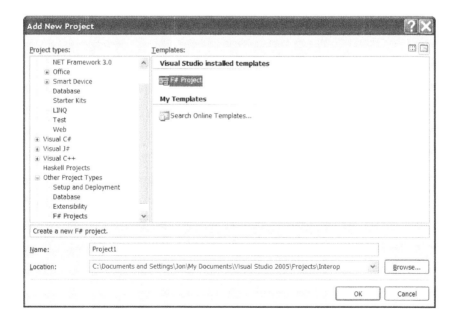

Figure 2.7   Creating a new F# project called `Project1` also inside the `Interop` solution.

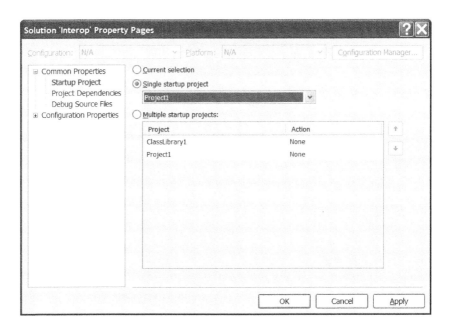

Figure 2.8   Setting the startup project of the Interop solution to the F# project Project1 rather than the C# project ClassLibrary1 as a DLL cannot be used to start an application.

For example, the following adds a member function foo() to the C# DLL:

```
using System;
using System.Collections.Generic;
using System.Text;

namespace ClassLibrary1
{
    public class Class1
    {
        public static void foo()
        {
            Console.WriteLine("foo");
        }
    }
}
```

The following F# source code adds the default location of the debug build of the C# DLL to the search path, references the DLL and then calls the foo() function:

```
#light
System.Environment.CurrentDirectory <-
    __SOURCE_DIRECTORY__
#I @"..\ClassLibrary1\bin\Debug"
```

```
#r "ClassLibrary1.dll"
ClassLibrary1.Class1.foo()
```

The syntax @" . . . " is used to quote a string without having to escape it manually.

This can be compiled or run interactively. When evaluated in an interactive solution, this F# code produces the following output:

```
> #light;;
> System.Environment.CurrentDirectory <-
    __SOURCE_DIRECTORY__;;
val it : unit = ()
> #I @"..\ClassLibrary1\bin\Debug";;
--> Added 'C:\Visual Studio 2005\Projects\Interop\
Project1\..\ClassLibrary1\bin\Debug' to library include
path
> #r "ClassLibrary1.dll";;
--> Referenced 'C:\Visual Studio 2005\Projects\Interop\
Project1\..\ClassLibrary1\bin\Debug\ClassLibrary1.dll'
> ClassLibrary1.Class1.foo();;
Binding session to 'C:\Visual Studio 2005\Projects\
Interop\Project1\..\ClassLibrary1\bin\Debug\
ClassLibrary1.dll'...
foo
val it : unit = ()
```

Note that the invocation of the C# function from F# prints "foo" as expected.

The same procedure can be used to add C# projects containing autogenerated code from the Windows Forms designer or by adding Web References to the C# project.

CHAPTER 3

DATA STRUCTURES

Scientific applications are among the most computationally intensive programs in existence. This places enormous emphasis on the efficiency of such programs. However, much time can be wasted by optimizing fundamentally inefficient algorithms and concentrating on low-level optimizations when much more productive higher-level optimizations remain to be exploited.

Too commonly, a given problem is shoe-horned into using arrays because more sophisticated data structures are prohibitively complicated to implement in many common languages. Examples of this problem, endemic in scientific computing, are rife. For example, Finite element materials simulations, numerical differential equation solvers, numerical integration, implicit surface tesselation and simulations of particle or fluid dynamics based around uniformly subdivided arrays when they should be based upon adaptively subdivided trees.

Occasionally, the poor performance of these inappropriately-optimized programs even drives the use of alternative (often approximate) techniques. Examples of this include the use of padding to round vector sizes up to integer-powers of two when computing numerical Fourier transforms (Fourier series). In order to combat this folklore-based approach to optimization, we shall introduce a more formal approach to quantifying the efficiency of computations. This approach is well known in computer science as *complexity theory*.

The single most important choice determining the efficiency of a program are the selection of algorithms and of data structures. Before delving into the broad spectrum of data structures accessible from F#, we shall study the notion of *algorithmic complexity*. This concept quantifies algorithm efficiency and, therefore, is essential for the objective selection of algorithms and data structures based upon their performance. Studying algorithmic complexity is the first step towards drastically improving program performance.

3.1 ALGORITHMIC COMPLEXITY

In order to compare the efficiencies of algorithms meaningfully, the time requirements of an algorithm must first be quantified. Although it is theoretically possible to predict the exact time taken to perform many operations such an approach quickly becomes intractable.

Consequently, exactness can be productively relinquished in favour of an approximate *but still quantitative* measure of the time taken for an algorithm to execute. This approximation, the conventional notion of algorithmic complexity, is derived as an upper- or lower-bound or average-case[10] of the amount of computation required, measured in units of some suitably-chosen primitive operations. Furthermore, asymptotic algorithmic efficiency is derived by considering these forms in the limit of infinite algorithmic complexity.

We shall begin by describing the notion of the primitive operations of an algorithm before deriving a mathematical description for the asymptotic complexity of an algorithm. Finally, we shall demonstrate the usefulness of algorithmic complexity in the optimization of a simple function.

3.1.1 Primitive operations

In order to derive an algorithmic complexity, it is necessary to begin by identifying some suitable primitive operations. The complexity of an algorithm is then measured as the total number of these primitive operations it performs. In order to obtain a complexity which reflects the time required to execute an algorithm, the primitive operations should ideally terminate after a constant amount of time. However, this restriction cannot be satisfied in practice (due to effectively-random interference from cache effects etc.), so primitive operations are typically chosen which terminate in a finite amount of time for any input, as close to a constant amount of time as possible.

For example, a first version of a function to raise a floating-point value x to a positive, integer power n may be implemented naïvely as:

```
> let rec ipow_1 x = function
```

---

[10] Average-case complexity is particularly useful when statistics are available on the likelihood of different inputs.

```
  | 0 -> 1.0
  | n -> x * ipow_1 x (n - 1);;
val ipow_1 : float -> int -> float
```

The `ipow_1` function executes an algorithm described by this recurrence relation:

$$x^n = \begin{cases} 1 & n = 0 \\ x \times x^{n-1} & \text{otherwise} \end{cases}$$

Consequently, this algorithm performs the floating-point multiply operation exactly n times in order to obtain its result, i.e. $x^0 = 1$, $x^1 = x \times 1$, $x^2 = x \times x \times 1$ and so on. Thus, floating-point multiplication is a logical choice of primitive operation. Moreover, this function multiplies finite-precision numbers and the algorithms used to perform this operation in practice (which are almost always implemented as dedicated hardware) always perform a *finite* number of more primitive operations at the bit level, regardless of their input. Thus, this choice of primitive operation will execute in a finite time regardless of its input.

We shall now examine an approximate but practically very useful measure of algorithmic complexity before exploiting this notion in the optimization of the `ipow_1` function.

3.1.2 Complexity

The complexity of an algorithm is the number of primitive operations it performs. For example, the complexity of the `ipow_1` function is $T(n) = n$.

As the complexity can be a strongly dependent function of the input, the mathematical derivation of the complexity quickly becomes intractable for reasonably complicated algorithms.

In practice, this is addressed in two different ways. The first approach is to derive the tightest possible bounds of the complexity. If such bounds cannot be obtained then the second approach is to derive bounds in the asymptotic limit of the complexity.

3.1.2.1 *Asymptotic complexity*   An easier-to-derive and still useful indicator of the performance of a function is its *asymptotic algorithmic complexity*. This gives the asymptotic performance of the function in the limit of infinite execution time.

Three notations exist for the asymptotic algorithmic complexity of a function $f(x)$:

$$\Omega(g(x)) \quad \Rightarrow \quad C_1 \leq \lim_{x \to \infty} \frac{f(x)}{g(x)}$$

$$O(g(x)) \quad \Rightarrow \quad \lim_{x \to \infty} \frac{f(x)}{g(x)} \leq C_2$$

$$\Theta(g(x)) \quad \Rightarrow \quad C_1 \leq \lim_{x \to \infty} \frac{f(x)}{g(x)} \leq C_2$$

for some constants $C_1, C_2 \in \mathbb{R}$.

The Θ form of asymptotic complexity is more restrictive and, therefore, conveys more information. In particular, "$f(x)$ is $\Theta(g(x))$" implies both "$f(x)$ is $\Omega(g(x))$" and "$f(x)$ is $O(g(x))$". The O notation is more commonly encountered as it represents the practically more important notion of the upper-bound of the complexity.

The formulation of the asymptotic complexity of a function leads to some simple but powerful manipulations:

- $f(x)$ is $O(a\,g(x))$, $a > 0 \Rightarrow f(n)$ is $O(g(x))$, i.e. constant prefactors can be removed.

- $f(x)$ is $O(x^a + x^b)$, $a > b > 0 \Rightarrow f(x)$ is $O(x^a)$, i.e. the polynomial term with the largest exponent dominates all other polynomial terms.

- $f(x)$ is $O(x^a + b^x)$, $a > 0, b > 0 \Rightarrow f(n)$ is $O(b^x)$, i.e. exponential terms dominate any polynomial terms.

- $f(x)$ is $O(a^x + b^x)$, $a > b > 0 \Rightarrow f(n)$ is $O(a^x)$, i.e. the exponential term a^x with the largest mantissa a dominates all other exponential terms.

These rules can be used to simplify an asymptotic complexity.

As the complexity of the `ipow_1` function is $T(n) = n$, the asymptotic complexities are clearly $O(n)$, $\Omega(n)$ and, therefore, $\Theta(n)$.

The algorithm behind the `ipow_1` function can be greatly improved upon by reducing the computation by a constant proportion at a time. In this case, this can be achieved by trying to halve n repeatedly, rather than decrementing it. The following recurrence relation describes such an approach:

$$x^n = \begin{cases} 1 & n = 0 \\ x & n = 1 \\ x(x^{\frac{1}{2}(n-1)})^2 & n > 1 \text{ and } n \text{ odd} \\ (x^{\frac{1}{2}n})^2 & n > 1 \text{ and } n \text{ even} \end{cases}$$

```
> let rec ipow_2 x = function
    | 0 -> 1.0
    | 1 -> x
    | n ->
        let x2 = ipow_2 x (n/2)
        x2 * if n mod 2 = 0 then x2 else x * x2;;
val ipow_2 : float -> int -> float
```

This variant is clearly more efficient as it avoids the re-computation of previous results, e.g. x^4 is factored into $(x^2)^2$ to use two floating-point multiplications instead of four. Quantifying exactly how much more efficient is suddenly a bit of a challenge!

We can begin by expanding the computation manually for some small n (see Table 3.1) as well as computing and plotting the exact number of integer multiplications performed for both algorithms as a function of n (shown in figure 3.1).

Table 3.1   Complexity of the `ipow_2` function measured as the number of multiply operations performed.

n	Computation	$T(n)$
0	1	0
1	x	0
2	$x \times x$	1
3	$x \times x^2$	2
4	$\left(x^2\right)^2$	2
5	$x \times \left(x^2\right)^2$	3
6	$\left(x \times x^2\right)^2$	3
7	$x \times \left(x \times x^2\right)^2$	4
8	$\left(\left(x^2\right)^2\right)^2$	3

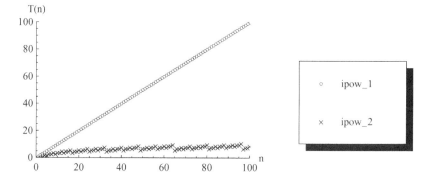

Figure 3.1   Complexities of the `ipow_1` and `ipow_2` functions in terms of the number $T(n)$ of multiplies performed.

Lower and upper bounds of the complexity can be derived by considering the minimum and maximum number of multiplies performed in the body of the `ipow_2` function, and the minimum and maximum depths of recursion.

The minimum number of multiplies performed in the body of a single call to the `ipow_2` function is 0 for $n \leq 1$, and 1 for $n > 1$. The function recursively halves n, giving a depth of recursion of 1 for $n \leq 1$, and at least $\lfloor \log_2 n \rfloor$ for $n > 1$. Thus, a lower bound of the complexity is 0 for $n \leq 1$, and $\log_2(n) - 1$ for $n > 1$.

The maximum number of multiplies performed in the body of a single call to the `ipow_2` function is 2. The depth of recursion is 1 for $n \leq 1$ and does not exceed $\lceil \log_2 n \rceil$ for $n > 1$. Thus, an upper bound of the complexity is 0 for $n \leq 1$, and $2(1 + \log_2 n)$ for $n > 1$.

From these lower and upper bounds, the asymptotic complexities of the `ipow_2` function are clearly $\Omega(\ln n)$, $O(\ln n)$ and, therefore, $\Theta(\ln n)$. The logarithmic complexity of `ipow_2` (illustrated in figure 3.2) originates from the divide-and-

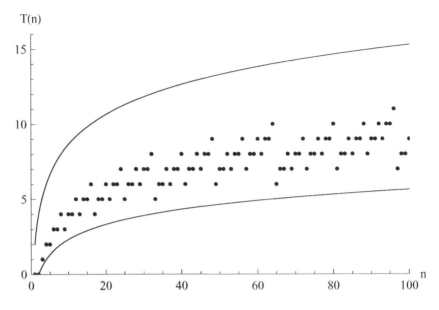

Figure 3.2   Complexities of the `ipow_2` function in terms of the number of multiplies performed, showing: exact complexity $T(n)$ (dots) and lower- and upper-bounds algorithmic complexities $\log_2(n) - 1 \leq T(n) \leq 2(1 + \log_2 n)$ for $n > 1$ (lines).

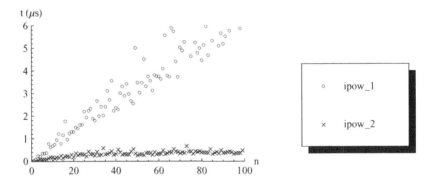

Figure 3.3   Measured performance of the `ipow_1` and `ipow_2` functions which have asymptotic algorithmic complexities of $\Theta(n)$ and $\Theta(\ln n)$, respectively.

conquer strategy, reducing the computation required by a constant factor (halving n) at each stage rather than by a constant absolute amount (decrementing n).

The actual performance of these two versions of the `ipow` function can be measured (see Figure 3.3). As expected from the algorithmic complexity, we find that the `ipow_2` function is considerably faster for large n.

| 0 | 1 | 2 | 3 | • • • | n-4 | n-3 | n-2 | n-1 |

Figure 3.4   Arrays are the simplest data structure, allowing fast, random access (reading or writing) to the ith element $\forall\, i \in \{0 \ldots n-1\}$ where n is the number of elements in the array. Elements cannot be added or removed without copying the whole array.

Asymptotic algorithmic complexity, as we have just described, should be considered *first* when trying to choose an efficient algorithm or data structure. On the basis of this, we shall now examine some of the wide variety of data structures accessible from F# in the context of the algorithmic complexities of operations over them.

3.2 ARRAYS

Of all the data structures, arrays will be the most familiar to the scientific programmer. Arrays have the following properties in F#:

- Mutable.

- Fixed size.

- Fast random access and length.

- Slow insertion, deletion, concatenation, partition and search.

More specifically, arrays are containers which allow the ith element to be extracted in $O(1)$ time (illustrated in figure 3.4). This makes them ideally suited in situations which require a container with fast random access. As the elements of arrays are typically stored contiguously in memory, they are often the most efficient container for iterating over the elements in order. This is the principal alluring feature which leads to their (over!) use in numerically intensive programs. However, many other data structures have asymptotically better performance on important operations.

3.2.1 Array literals

As we have already seen, the F# language provides a notation for describing arrays:

```
> let a = [|1; 2|]
  let b = [|3; 4; 5|]
  let c = [|6; 7; 9|];;
val a : int array
val b : int array
val c : int array
```

In F#, arrays are mutable, meaning that the elements in an array can be altered in-place.

3.2.2 Array indexing

The element at index i of an array b may be read using the short-hand syntax b. [i] :

```
> b.[1];;
val it : int = 4
```

Note that array indices run from $\{0 \ldots n-1\}$ for an array containing n elements.
Array elements may be set using the syntax used for mutable record fields, namely:

```
> c.[2] <- 8;;
val it : unit = ()
```

The contents of the array c have now been altered:

```
> c;;
val it : int array = [|6; 7; 8|]
```

Any attempt to access an array element which is outside the bounds of the array
results in an exception being raised at run-time:

```
> c.[3] <- 8;;
System.IndexOutOfRangeException: Index was outside the
bounds of the array
```

The mutability of arrays typically leads to the use of an imperative style when
arrays are being used.

The core F# library provides several functions which act upon arrays in the Array
module. We shall examine some of these functions before looking at the more exotic
array functions offered by F#.

3.2.3 Array concatenation

The append function concatenates two arrays:

```
> Array.append a b;;
val it : int array = [|1; 2; 3; 4; 5|]
```

The append function has complexity $\Theta(n)$ where n is the length of the resulting
array.

The concat function concatenates a list of arrays:

```
> let e = Array.concat [a; b; c];;
val e : int array
> e;;
val it : int array = [|1; 2; 3; 4; 5; 6; 7; 8|]
```

The concat function has complexity $\Theta(n+m)$ where n is the length of the
resulting array and m is the length of the supplied list.

3.2.4 Aliasing

As we discussed in section 1.6.1, values are referenced so a new variable created from an existing array refers to the existing array. Thus the complexity of creating a new variable which refers to an existing array is $\Theta(1)$, i.e. independent of the number of elements in the array. However, all alterations to the array are visible from any variables which refer to the array. For example, the following creates a variable called d which refers to the same array as the variable called c:

```
> let d = c;;
val d : int array = [|6; 7; 8|]
```

The effect of altering the array via either c or d can be seen from both c and d, i.e. they are the same array:

```
> d.[0] <- 17;;
val it : unit = ()
> c, d;;
val it : int array * int array =
  ([|17; 7; 8|], [|17; 7; 8|])
> c.[0] <- 6;;
val it : unit = ()
> c, d;;
val it : int array * int array =
  ([|6; 7; 8|], [|6; 7; 8|])
```

The copy function returns a new array which contains the same elements as the given array. For example, the following creates a variable d (superseding the previous d) which is a copy of c:

```
> let d = Array.copy c;;
val d : int array
```

Altering the elements of the copied array d does not alter the elements of the original array c:

```
> d.[0] <- 17;;
val it : unit = ()
> c, d;;
val it : int array * int array =
  ([|6; 7; 8|], [|17; 7; 8|])
```

Aliasing can sometimes be productively exploited in heavily imperative code. However, aliasing is often a source of problems because the sharing of mutable data structures quickly gets confusing. This is why F# encourages a purely functional style of programming using only immutable but performance sometimes leads to the use of mutable data structures because they are more difficult to use correctly but are often faster, particularly for single-threaded use.

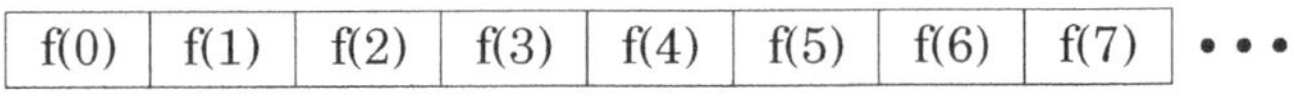

f(0)	f(1)	f(2)	f(3)	f(4)	f(5)	f(6)	f(7)	• • •

Figure 3.5   The higher-order `Array.init` function creates an array $a_i = f(i)$ for $i \in \{0 \ldots n - 1\}$ using the given function f.

3.2.5 Subarrays

The `sub` function returns a new array which contains a copy of the range of elements specified by a starting index and a length. For example, the following copies a sub-array of 5 elements starting at index 2 (the third element):

```
> Array.sub e 2 5;;
val it : int array = [|3; 4; 5; 6; 7|]
```

In addition to these conventional array functions, F# offers some more exotic functions. We shall now examine these functions in more detail.

3.2.6 Creation

The higher-order `init` function creates an array, filling the elements with the result of applying the given function to the index $i \in \{0 \ldots n - 1\}$ of each element (illustrated in figure 3.5). For example, the following creates the array $a_i = i^2$ for $i \in \{0 \ldots 3\}$:

```
> let a = Array.init 4 (fun i -> i * i);;
val a : int array

> a;;
val it : int array = [|0; 1; 4; 9|]
```

Several higher-order functions are provided to allow functions to be applied to the elements of arrays in various patterns.

3.2.7 Iteration

The higher-order function `iter` executes a given function on each element in the given array in turn and returns the value of type `unit`. The purpose of the function passed to this higher-order function must, therefore, lie in any side-effects it incurs. Hence, the `iter` function is only of use in the context of imperative, and not functional, programming. For example, the following prints the elements of an `int` array:

```
> Array.iter (printf "%d\n") a;;
0
1
4
9
```

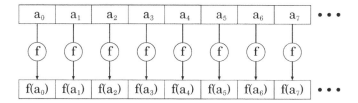

Figure 3.6   The higher-order `Array.map` function creates an array containing the result of applying the given function f to each element in the given array a.

Functions like `iter` are sometimes referred to as aggregate operators. Several such functions are often applied in sequence, in which case the pipe operator `|>` can be used to improve clarity. For example, the above could be written equivalently as:

```
> a |> Array.iter (printf "%d\n");;
0
1
4
9
```

The `iter` function is probably the simplest of the higher-order functions.

3.2.8 Map

The map function applies a given function to each element in the given array, returning an array containing each result (illustrated in figure 3.6). For example, the following creates the array $b_i = a_i^2$:

```
> let b = Array.map (fun e -> e * e) a;;
val b : int array
> b;;
val b : int array = [|0; 1; 16; 81|]
```

Higher-order map functions transform data structures into other data structures of the same kind. Higher-order functions can also be used to compute other values from data structures, such as computing a scalar from an array.

3.2.9 Folds

The higher-order `fold_left` and `fold_right` functions accumulate the result of applying their function arguments to each element in turn. The `fold_left` function is illustrated in figure 3.7.

In the simplest case, the fold functions can be used to accumulate the sum or product of the elements of an array by folding the addition or multiplication operators over the elements of the array respectively, starting with a suitable base case. For example, the sum of 3, 4, and 5 can be calculated using:

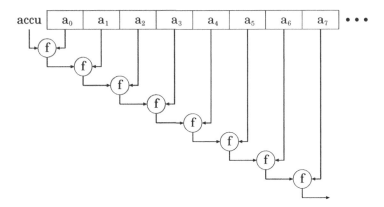

Figure 3.7   The higher-order `Array.fold_left` function repeatedly applies the given function f to the current accumulator and the current array element to produce a new accumulator to be applied with the next array element.

```
> Array.fold_left ( + ) 0 [|3; 4; 5|];;
val it : int = 98
```

We have already encountered this functionality in the context of the `fold_range` function developed in section 2.1. However, arrays may contain arbitrary data of arbitrary types (polymorphism) whereas the `fold_range` function was only applicable to sequences of consecutive `int`s.

For example, an array could be converted to a list by prepending elements using the cons operator:

```
> let to_list a =
    Array.fold_right (fun h t -> h :: t) a [];;
val to_list : 'a array -> 'a list
```

Applying this function to an example array produces a list with the elements of the array:

```
> to_list [|0; 1; 4; 9|];;
val it : int list = [0; 1; 4; 9]
```

This `to_list` function uses `fold_right` to cumulatively prepend to the list t each element h of the array a in reverse order, starting with the base-case of an empty list `[]`. The result is a list containing the elements of the array. The `to_list` function is, in fact, already in the `Array` module:

```
> Array.to_list [|0; 1; 4; 9|];;
val it : int list = [0; 1; 4; 9]
```

Although slightly more complicated than `iter` and `map`, the `fold_left` and `fold_right` functions are very useful because they can produce results of any type, including accumulated primitive types as well as different data structures.

3.2.10 Sorting

The contents of an array may be sorted in-place using the higher-order `Array.sort` function, into an order specified by a given total order function (the predefined `compare` function, in this case):

```
> let a = [|1; 5; 3; 4; 7; 9|];;
val a : int array
> Array.sort compare a;;
val it : unit = ()
> a;;
val it : int array = [|1; 3; 4; 5; 7; 9|]
```

In addition to these higher-order functions, arrays can be used in pattern matches.

3.2.11 Pattern matching

When pattern matching, the contents of arrays can be used in patterns. For example, the vector cross product $\mathbf{a} \times \mathbf{b}$ could be written:

```
> let cross (a : float array) b =
    match a, b with
    | [|x1; y1; z1|], [|x2; y2; z2|] ->
        [|y1 * z2 - z1 * y2;
          z1 * x2 - x1 * z2;
          x1 * y2 - y1 * x2|]
    | _ -> invalid_arg "cross";;
val cross : float array -> float array -> float array
```

Note the use of a type annotation on the first argument to ensure that the function applies to `float` elements rather than the default `int` elements.

Thus, patterns over arrays can be used to test the number and value of all elements, forming a useful complement to the map and fold algorithms.

3.3 LISTS

Lists have the following properties:

- Immutable.

- Fast prepend and decapitation (see figure 3.8).

- Slow random access, length, insertion, deletion, concatenation, partition and search.

Arguably the simplest and most commonly used data structure, lists allow two fundamental operations. The first is decapitation of a list into two parts, the *head* (the

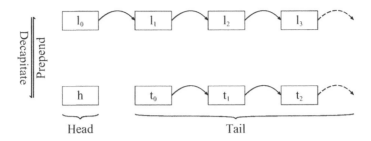

Figure 3.8   Lists are the simplest, arbitrarily-extensible data structure. Decapitation splits a list l_i $i \in \{0 \ldots n-1\}$ into the head element h and the tail list t_i $i \in \{0 \ldots n-2\}$.

first element of the list) and the *tail* (a list containing the remaining elements). The second is the reverse operation of prepending an element onto a list to create a new list. The complexities of both operations are $\Theta(1)$, i.e. the time taken to perform these operations is independent of the number of elements in the list. Thus, lists are ideally suited for the creation of a data structure containing an unknown number of elements (such as the loading of an arbitrarily-long sequence of numbers).

As we have already seen, the F# language implements lists natively. In particular, the cons operator `::` can be used both to prepend an element and to represent decapitated lists in patterns. Unlike arrays, the implementation of lists is functional, so operations on lists produce new lists.

The `List` module contains the `iter`, `map`, `fold_left` and `fold_right` functions, equivalent to those for `arrays`, a `flatten` function, which provides equivalent functionality to that of `Array.concat` (i.e. to concatenate a list of lists into a single list). In particular, the `append` function has the pseudonym `@`:

```
> [1; 2] @ [3; 4];;
val it : int list = [1; 2; 3; 4]
```

The `List` module also contains several functions for sorting and searching.

3.3.1 Sorting

The contents of a list may be sorted using the higher-order `List.sort` function, into an order specified by a given total order function (the predefined `compare` function, in this case):

```
> List.sort compare [1; 5; 3; 4; 7; 9];;
val it : int list = [1; 3; 4; 5; 7; 9]
```

Lists can be searched using a variety of functions.

3.3.2 Searching

There are three main forms of searching provided by the `List` module.

3.3.2.1 *Membership* An element may be tested for membership in a list using the List.mem function:

```
> List.mem 4 [1; 3; 4; 5; 7; 9];;
val it : bool = true
> List.mem 6 [1; 3; 4; 5; 7; 9];;
val it : bool = false
```

This is the simplest form of searching that can be applied to lists and is an easy but inefficient way to handle sets of values.

3.3.2.2 *Predicate* The first element in a list that matches a given predicate function may be extracted using the higher-order List.find function:

```
> List.find (fun i -> (i-6)*i > 0) [1; 3; 4; 5; 7; 9];;
val it : int = 7
```

This function raises an exception if all the elements in the given list fail to match the given predicate:

```
> List.find (fun i -> (i-6)*i = 0) [1; 3; 4; 5; 7; 9];;
System.IndexOutOfRangeException: Index was outside the
bounds of the array
```

The related function List.find_all returns the list of all elements that match the given predicate function:

```
> [1; 3; 4; 5; 7; 9]
   |> List.find_all (fun i -> (i-6)*i > 0);;
val it : int list = [7; 9]
```

Searching lists using predicate functions works well when several different predicates are used and performance is unimportant.

3.3.2.3 *Association lists* The contents of a list of key-value pairs may be searched using the List.assoc function to find the value corresponding to the first matching key. For example, the following list list contains (i, i^2) key-value pairs:

```
> let list =
    [ for i in 1 .. 5 ->
        i, i*i ];;
val list : (int * int) list
> list;;
val it : (int * int) list
  = [(1, 1); (2, 4); (3, 9); (4, 16); (5, 25)]
```

Searching list for the key $i = 4$ using the List.assoc function finds the corresponding value $i^2 = 16$:

```
> List.assoc 4 list;;
```

```
val it : int = 16
```

As we have seen, lists are easily searched in a variety of different ways. Consequently, lists are one of the most useful generic containers provided by the F# language. However, testing for membership is really a set-theoretic operation and the built-in HashSet and Set modules provide asymptotically more efficient ways to implement sets as well as extra functionality. Similarly, the ability to map keys to values in an association list is provided more efficiently by hash tables and maps. All of these data structures are discussed later in this chapter.

3.3.3 Filtering

The ability to grow lists makes them ideal for filtering operations based upon arbitrary predicate functions. The List.partition function splits a given list into two lists containing those elements which match the predicate and those which do not. The following example uses the predicate $x_i \leq 3$:

```
> List.partition (fun x -> x <= 3) [1 .. 5];;
val it : int list * int list = ([1; 2; 3], [4; 5])
```

Similarly, the List.filter function returns a list containing only those elements which matched the predicate, i.e. the first list that List.partition would have returned:

```
> List.filter (fun x -> x <= 3) [1 .. 5];;
val it : int list = [1; 2; 3]
```

The partition and filter functions are ideally suited to arbitrarily extensible data structures such as lists because the length of the output(s) cannot be precalculated.

3.3.4 Maps and folds

In addition to the conventional higher-order iter, map, folds, sorting and searching functions, the List module contains several functions which act upon pairs of lists. These functions all assume the lists to be of equal length. If they are found to be of different lengths then an Invalid_argument exception is raised. We shall now elucidate these functions using examples from vector algebra.

The higher-order function map2 applies a given function to each pair of elements from two equal-length lists, producing a single list containing the results. The type of the map2 function is:

```
val map2 :
    ('a -> 'b -> 'c) -> 'a list -> 'b list -> 'c list
```

The map2 function can be used to write a function to convert a pair of lists into a list of pairs:

```
> let list_combine a b =
    List.map2 (fun a b -> a, b) a b;;
```

```
val list_combine : 'a list -> 'b list -> ('a, 'b) list
```

Applying the list_combine function to a pair of lists of equal lengths combines them into a list of pairs:

```
> list_combine [1; 2; 3] ["a"; "b"; "c"];;
val it : (int * string) list =
  [(1, "a"); (2, "b"); (3, "c")]
```

Applying the list_combine function to a pair of lists of unequal lengths causes an exception to be raised by the map2 function:

```
> list_combine [1; 2; 3] [2; 3; 4; 5];;
Microsoft.FSharp.InvalidArgumentException: map2
```

In fact, the functionality of this list_combine function is already provided by the combine function in the List module of the F# standard library.

Vector addition for vectors represented by lists can be written in terms of the map2 function:

```
> let vec_add (a : float list) b =
    List.map2 ( + ) a b;;
val vec_add : float list -> float list -> float list
```

When given a pair of lists a and b of floating-point numbers, this function creates a list containing the sum of each corresponding pair of elements from the two given lists, i.e. a + b:

```
> vec_add [1.0; 2.0; 3.0] [2.0; 3.0; 4.0];;
val it : float list = [3.0; 5.0; 7.0]
```

The higher-order fold_left2 and fold_right2 functions in the List module are similar to the fold_left and fold_right functions, except than they act upon two lists simultaneously instead of one. The types of these functions are:

```
val fold_left2 :
  ('a -> 'b -> 'c -> 'a) -> 'a -> 'b list -> 'c list ->
    'a
val fold_right2 :
  ('a -> 'b -> 'c -> 'c) -> 'a list -> 'b list -> 'c ->
    'c
```

Thus, the fold_left2 and fold_right2 functions can be used to implement many algorithms which consider each pair of elements in a pair of lists in turn. For example, the vector dot product could be written succinctly using fold_left2 by accumulating the products of element pairs from the two lists:

```
> let vec_dot (a : float list) (b : float list) =
    List.fold_left2 (fun d a b -> d + a * b) 0.0 a b;;
val vec_dot : float list -> float list -> float
```

When given two lists, a and b, of floating-point numbers, this function accumulates the products $a_i \times b_i$ of each pair of elements from the two given lists, i.e. the vector dot product $\mathbf{a} \cdot \mathbf{b}$:

```
> vec_dot [1.0; 2.0; 3.0] [2.0; 3.0; 4.0];;
val it : float = 20.0
```

The ability to write such functions using maps and folds is clearly an advantage in the context of scientific computing. Moreover, this style of programming can be seamlessly converted to using much more exotic data structures, as we shall see later in this chapter. In some cases, algorithms over lists cannot be expressed easily in terms of maps and folds. In such cases, pattern matching can be used instead.

3.3.5 Pattern matching

Patterns over lists can not only reflect the number and value of all elements, as they can in arrays, but can also be used to test initial elements in the list using the cons operator : : to decapitate the list. In particular, pattern matching can be used to examine sequences of list elements. For example, the following function "downsamples" a signal, represented as a list of floating-point numbers, by averaging pairs of elements:

```
> let rec downsample : float list -> float list =
    function
    | [] -> []
    | h1::h2::t -> 0.5 * (h1 + h2) :: downsample t
    | [_] -> invalid_arg "downsample";;
val downsample : float list -> float list
```

This is a simple, recursive function which uses a pattern match containing three patterns. The first pattern downsamples the empty list to the empty list. This pattern acts as the base-case for the recursive calls of the function (equivalent to the base-case of a recurrence relation). The second pattern matches the first two elements in the list (h1 and h2) and the remainder of the list (t). Matching this pattern results in prepending the average of h1 and h2 onto the list resulting from downsampling the remaining list t. The third pattern matches a list containing any single element, raising an exception if this erroneous input is encountered:

```
> downsample [5.0];;
Exception: Invalid_argument "downsample".
```

As these three patterns are completely distinct (any input list necessarily matches one and only one pattern) they could, equivalently, have been presented in any order in the pattern match.

The downsample function can be used to downsample an eight-element list into a four-element list by averaging pairs of elements:

```
> [0.0; 1.0; 0.0; -1.0; 0.0; 1.0; 0.0; -1.0]
  |> downsample;;
```

```
val it : float list = [0.5; -0.5; 0.5; -0.5]
```

The ability to perform pattern matches over lists is extremely useful, resulting in a very concise syntax for many operations which act upon lists.

Note that, in the context of lists, the `iter` and `map` higher-order functions can be expressed succinctly in terms of folds:

```
let iter f list =
  List.fold_left (fun () e -> f e) () list
let map f list =
  List.fold_right (fun h t -> f h :: t) list []
```

and that all of these functions can be expressed, albeit more verbosely, using pattern matching. The `iter` function simply applies the given function `f` to the head `h` of the list and then recurses to iterate over the element in the tail `t`:

```
let rec iter f = function
  | [] -> ()
  | h::t -> f h; iter f t
```

The `map` function applies the given function `f` to the head `h` of the list, prepending the result `f h` onto the result `map f t` of recursively mapping over the elements in the tail `t`:

```
let rec map f = function
  | [] -> []
  | h::t -> f h :: map f t
```

The `fold_left` function applies the given function `f` to the current accumulator `accu` and the head `h` of the list, passing the result as the accumulator for folding over the remaining elements in the tail `t` of the list:

```
let rec fold_left f accu list =
  match list with
  | h::t -> fold_left f (f accu h) t
  | [] -> accu
```

The `fold_right` function applies the given function `f` to the head `h` of the list and the result of recursively folding over the remaining elements in the tail `t` of the list:

```
let rec fold_right f list accu =
  match list with
  | h::t -> f h (fold_right f t accu)
  | [] -> accu
```

Thus, the `map` and `fold` functions can be thought of as higher-order functions which have been factored out of many algorithms. In the context of scientific programming, factoring out higher-order functions can greatly increase clarity, often providing new insights into the algorithms themselves. In chapter 6, we shall pursue this, developing several functions which supplement those in the standard library.

Having examined the two simplest containers provided by the F# language itself, we shall now examine some more sophisticated containers which are provided in the standard library.

3.4   SETS

Sets have the following properties:

- Immutable.

- Sorted.

- Duplicates are removed.

- Fast insertion, deletion, membership, partition, union, difference and intersection.

- No random access.

In the context of data structures, the term "set" typically means a sorted, unique, associative container. Sets are "sorted" containers because the elements in a set are stored in order according to a given comparison function. Sets are "unique" containers because they do not duplicate elements (adding an existing element to a set results in the same set). Sets are "associative" containers because elements determine how they are stored (using the specified comparison function).

The F# standard library provides sets which are implemented as balanced binary trees[11]. This allows a single element to be added or removed from a set containing n elements in $O(\ln n)$ time. Moreover, the F# implementation also provides functions `union`, `inter` and `diff` for performing the set-theoretic operations union, intersection and difference efficienctly.

In order to implement the set-theoretic operations between sets correctly, the sets used must be based upon the same comparison function. If no comparison function is specified then the built-in `compare` function is used by default. Sets implemented using custom comparison functions can be created using the `Set.Make` function.

3.4.1   Creation

The empty set can be obtained as:

```
> Set.empty;;
val it : Set<'a>
```

The type `Set<'a>` represents a set of values of type `'a`.

A set containing a single element is called a *singleton* set and can be created using the `Set.singleton` function:

[11]Balanced binary trees will be discussed in more detail later in section 3.10.

```
> let set1 = Set.singleton 3;;
val set1 : Set<int>
```

Sets can also be created from sequences using the set function:

```
> set [1; 2; 3];;
val it : Set<int> = seq [1; 2; 3]
```

Once created, sets can have elements added and removed.

3.4.2 Insertion

As sets are implemented in a functional style, adding an element to a set returns a new set that shares immutable data with the old set:

```
> let set2 = Set.add 5 set1;;
val set2 : Set<int>
```

Both the old set and the new set are still valid:

```
> set1, set2;;
val it : Set<int> * Set<int> = (seq [3], seq [3; 5])
```

Sets remove duplicates:

```
> let s = set [10; 1; 9; 2; 8; 4; 7; 4; 6; 7; 7];;
val s : Set<int>
```

A set may be converted into other data structures, such as a list using the `to_list` function in the `Set` module:

```
> Set.to_list s;;
val it : int list = [1; 2; 4; 6; 7; 8; 9; 10]
```

As we shall see in section 3.8, sets are also a kind of sequence.

3.4.3 Cardinality

The number of elements in a set, known as the *cardinality* of the set, is given by:

```
> Set.cardinal s;;
val it : int = 8
```

Perhaps the most obvious functionality of sets is that of set-theoretic operations.

3.4.4 Set-theoretic operations

We can also demonstrate the set-theoretic union, intersection and difference operations. For example:

$$\{1,3,5\} \cup \{3,5,7\} = \{1,3,5,7\}$$

```
> Set.union (set [1; 3; 5]) (set [3; 5; 7]);;
val it : Set<int> = [1; 3; 5; 7]
```

$$\{1,3,5\} \cap \{3,5,7\} = \{3,5\}$$

```
> Set.inter (set [1; 3; 5]) (set [3; 5; 7]);;
val it : Set<int> = [3; 5]
```

$$\{1,3,5\} \setminus \{3,5,7\} = \{1\}$$

```
> Set.diff (set [1; 3; 5]) (set [3; 5; 7]);;
val it : Set<int> = [1]
```

Set union and difference can also be obtained by applying the + and - operators:

```
> set [1; 3; 5] + set [3; 5; 7];;
val it : Set<int> = [1; 3; 5; 7]
> set [1; 3; 5] - set [3; 5; 7];;
val it : Set<int> = [1]
```

The subset function tests if $A \subset B$. For example, $\{4,5,6\} \subset \{1 \ldots 10\}$:

```
> Set.subset (set [2; 4; 6]) s;;
val it : bool = true
```

Of the data structures examined so far (lists, arrays and sets), only sets are not concrete data structures. Specifically, the underlying representation of sets as balanced binary trees is completely abstracted away from the user. In some languages, this can cause problems with notions such as equality but, in F#, sets can be compared safely using the built-in comparison operators and function.

3.4.5 Comparison

Although sets are non-trivial data structures internally, the polymorphic comparison functions (<, <=, =, >=, >, <> and compare) can be applied to sets and data structures containing sets in F#:

```
> Set.of_list [1; 2; 3; 4; 5] =
    Set.of_list [5; 4; 3; 2; 1];;
val it : bool = true
```

In chapter 12, we shall use a set data structure to compute the set of nth-nearest neighbours in a graph and apply this to atomic-neighbour computations on a simulated molecule. In the meantime, we have more data structures to discover.

3.5 HASH TABLES

Hash tables have the following properties:

- Mutable.

- Each key maps onto a list of values.

- Fast insertion, deletion and search.

A hash table is an associative container mapping *keys* to corresponding *values*. We shall refer to the key-value pairs stored in a hash table as *bindings* or elements.

In terms of utility, hash tables are an efficient way to implement a mapping from one kind of value to another. For example, to map strings onto functions. In order to provide their functionality, hash tables provide `add`, `replace` and `remove` functions to insert, replace and delete bindings, respectively, a `find` function to return the first value corresponding to a given key and a `find_all` function to return all corresponding values.

Internally, hash tables compute an integer value, known as a *hash*, from each given key. This hash of a key is used as an index into an array in order to find the value corresponding to the key. The hash is computed from the key such that two identical keys share the same hash and two different keys are likely (but not guaranteed) to produce different hashes. Moreover, hash computation is restricted to $\Theta(1)$ time complexity, typically by terminating if a maximum number of computations is reached. Assuming that no two keys in a hash table produce the same hash, finding the value corresponding to a given key takes $\Theta(1)$ time[12].

The F# standard library contains an imperative implementation of hash tables in the `Hashtbl` module.

3.5.1 Creation

The `Hashtbl.of_seq` function can be used to create hash tables from association sequences:

```
> let m =
    ["Hydrogen", 1.0079; "Carbon", 12.011;
     "Nitrogen", 14.00674; "Oxygen", 15.9994;
     "Sulphur", 32.06]
    |> Hashtbl.of_seq;;
val m : Hashtbl.HashTable<string, float>
```

The resulting hash table m, of type `Hashtbl.HashTable<string, float>`, represents the following mapping from strings to floating-point values:

$$
\begin{array}{rcl}
\text{Hydrogen} & \rightarrow & 1.0079 \\
\text{Carbon} & \rightarrow & 12.011 \\
\text{Nitrogen} & \rightarrow & 14.00674 \\
\text{Oxygen} & \rightarrow & 15.9994 \\
\text{Sulphur} & \rightarrow & 32.06
\end{array}
$$

Hash tables are primarily used for searching.

---

[12]Computing the hash in $\Theta(1)$ time and then using it to access an array element, also in $\Theta(1)$ time.

3.5.2 Searching

Having been filled at run-time, the hash table may be used to look-up the values corresponding to given keys. For example, we can find the average atomic weight of carbon:

```
> Hashtbl.find m "Carbon";;
val it : float = 12.011
```

Looking up values in a hash table is such a common operation that a more concise syntax for this operation has been provided via the getter of the Item member, accessed using the notation:

m.[*key*]

For example:

```
> m.["Carbon"];;
val it : float = 12.011
```

Hash tables also allow bindings to be added and removed.

3.5.3 Insertion, replacement and removal

A mapping can be added using the Hashtbl.add function:

```
> Hashtbl.add m "Tantalum" 180.9;;
val it : unit = ()
```

Note that adding a new key-value binding shadows any existing bindings rather than replacing them. Subsequently removing such a binding makes the previous binding for the same key visible again if there was one.

Bindings may be replaced using the Hashtbl.replace function:

```
> Hashtbl.replace m "Tantalum" 180.948;;
val it : unit = ()
```

Any previous binding with the same key is then replaced.

The setter of the Item property is also implemented, allowing the following syntax to be used to replace a mapping in a hash table m:

m.[*key*] <- *value*

If the hash table did not contain the given key then a new key is added.

For example, the following inserts a mapping for tantalum:

```
> m.["Tantalum"] <- 180.948;;
val it : unit = ()
```

If necessary, we can also delete mappings from the hash table, such as the mapping for Oxygen:

```
> Hashtbl.remove m "Oxygen";;
val it : unit = ()
```

The usual higher-order functions are also provided for hash tables.

3.5.4   Higher-order functions

The remaining mappings in the hash table are most easily printed using the `iter` function in the `Hashtbl` module:

```
> Hashtbl.iter (printf "%s -> %f\n") m;;
Carbon -> 12.011
Nitrogen -> 14.00674
Sulphur -> 32.06
Hydrogen -> 1.0079
Tantalum -> 180.948
```

Note that the order in which the mappings are supplied by `Hashtbl.iter` (and `map` and `fold`) is effectively random. In fact, the order is related to the hash function.

Hash tables can clearly be useful in the context of scientific programming. However, a functional alternative to these imperative hash tables can sometimes be desirable. We shall now examine a functional data structure which uses different techniques to implement the same functionality of mapping keys to corresponding values.

3.6   MAPS

Maps have the following properties:

- Immutable.

- Each key maps to a single value.

- Fast insertion, deletion and search.

We described the functional implementation of the set data structure provided by F# in section 3.4. The core F# library provides a similar data structure known simply as a map[13].

Much like a hash table, the map data structure associates keys with corresponding values. Consequently, the map data structure also provides `add` and `remove` functions to insert and delete mappings, respectively, and a `find` function which returns the value corresponding to a given key.

Unlike hash tables, maps are represented internally by a balanced binary tree, rather than an array, and maps differentiate between keys using a specified total ordering function, rather than a hash function.

Due to their design differences, maps have the following advantages over hash tables:

- Immutable: operations on maps derive new maps without invalidating old maps.

---

[13]Not to be confused with the higher-order map function provided with many data structures.

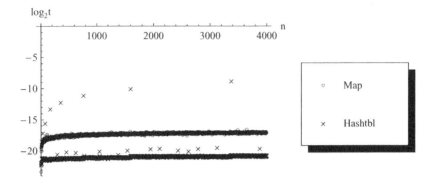

Figure 3.9   Measured performance (time t in seconds) for inserting key-value pairs into hash tables and functional maps containing $n-1$ elements. Although the hash table implementation results in better average-case performance, the $O(n)$ time-complexity incurred when the hash table is resized internally produces much slower worst-case performance by the hash table.

- Stable $O(\log n)$ time-complexity for inserting and removing a mapping, compared to unstable, amortized $\Theta(1)$ time-complexity in the case of hash tables (which may take up to $O(n)$ for some insertions, as illustrated in figure 3.9).

- Maps are based upon comparison and hash tables are based upon hashing. Comparisons are easier to define and more readily available.

- If comparison is faster than hashing (e.g. for structurally large keys, where comparison can return early) then a map may be faster than a hash table.

- The key-value pairs in a map are kept sorted by key.

However, maps also have the following disadvantages compared to hash tables:

- Logarithmic $O(\ln n)$ time-complexity for insertion, lookup and removal, compared to amortized $\Theta(1)$ time-complexity in the case of hash tables (see figure 3.9).

- If comparison is slower than hashing (e.g. for small keys) then a hash table is likely to be faster than a map.

In single threaded applications, a map is typically $10\times$ slower than a hash table in practice.

3.6.1   Creation

We shall now demonstrate the functionality of the map data structure reusing the example of mapping strings to floating-point values. A map data structure containing no mappings is represented by the value:

```
> let m = Map.empty;;
val m : Map<'a, 'b>
```

Note that this value is polymorphic, allowing the empty map to be used as the basis of mappings between any types.

Mappings may be added to m using the `Map.add` function. As a functional data structure, adding elements to a map returns a map containing both the new and old mappings. Hence we repeatedly supersede the old data structure m with a new data structure m:

A map may be built from a list using the `Map.of_list` function:

```
> let m =
    Map.of_list
    ["Hydrogen", 1.0079; "Carbon", 12.011;
     "Nitrogen", 14.00674; "Oxygen", 15.9994;
     "Sulphur", 32.06]
val m : Map<string, float>
```

Note that the type has been inferred to be a mapping from strings to floats.

The contents of a map can be converted to a list using the `fold` function:

```
> let list_of map =
    Map.fold (fun h t -> h :: t) map [];;
val list_of : Map<'a, 'b> -> ('a, 'b) list
```

Once created, a map can be searched.

3.6.2 Searching

We can use the map m to find the average atomic weight of carbon:

```
> Map.find "Carbon" m;;
val it : float = 12.011
```

The syntax used for hash table lookup can also be used on maps:

```
> m.["Carbon"];;
val it : float = 12.011
```

However, as maps are immutable they cannot be altered in-place and, consequently, the `Item` setter method is not implemented. Adding a binding to a map produces a new immutable map that shares much of its internals with the old map (which remains valid). For example, we can create a new map with an additional binding:

```
> let m2 = Map.add "Tantalum" 180.948 m;;
val m2 : Map<string, float>
```

Both the old map m and the new map m2 are valid:

```
> list_of m, list_of m2;;
val it : (string * float) list * (string * float) list
```

```
= ([[("Carbon", 12.011); ("Hydrogen", 1.0079);
     ("Nitrogen", 14.00674); ("Oxygen", 15.9994);
     ("Sulphur", 32.06)],
    [("Carbon", 12.011); ("Hydrogen", 1.0079);
     ("Nitrogen", 14.00674); ("Oxygen", 15.9994);
     ("Sulphur", 32.06); ("Tantalum", 180.948)])]
```

Note that the key-value pairs are sorted by key, alphabetically in this case.

Deleting mappings from the functional map data structure produces a new data structure containing most of the old data structure:

```
> Map.remove "Oxygen" m2;;
val it : Map<string, float>
  = [("Carbon", 12.011); ("Hydrogen", 1.0079);
     ("Nitrogen", 14.00674); ("Sulphur", 32.06);
     ("Tantalum", 180.948)]
```

Maps contain the usual higher-order functions.

3.6.3 Higher-order functions

The remaining mappings are most easily printed by iterating a print function over the key-value pairs:

```
> Map.iter (printf "%s -> %f\n") m2;;
Carbon -> 12.011
Hydrogen -> 1.0079
Nitrogen -> 14.00674
Oxygen -> 15.9994
Sulphur -> 32.06
Tantalum -> 180.948
```

Note that m2 still contains the entry for oxygen as deleting this binding created a new map that was discarded.

The ability to evolve the contents of data structures along many different lineages during the execution of a program can be very useful. This is much easier when the data structure provides a functional, rather than an imperative, interface. In an imperative style, such approaches would most likely involve the inefficiency of explicitly duplicating the data structure (e.g. using the Hashtbl.copy function) at each fork in its evolution. In contrast, functional data structures provide this functionality naturally and, in particular, will share data between lineages.

Having examined the data structures provided with F#, we shall now summarise the relative advantages and disadvantages of these data structures using the notion of algorithmic complexity developed in section 3.1.

Table 3.2   Functions implementing common operations over data structures.

	Array	List	Set	Hash table	Map
Create	init	init	of_array	of_list	of_list
Insert	-	-	add	replace	add
Search	find	find	mem	find	find
Remove	-	remove_assoc	remove	remove	remove
Sort	sort	sort	N/A	N/A	N/A
Index	get	nth/assoc	N/A	find	find

Table 3.3   Asymptotic algorithmic complexities of operations over different data structures containing n elements where $i \in \{0 \ldots n - 1\}$ is a parameter of some of the operations, e.g. insert at index i.

	Array	List	Set	Hash table	Map
Create	$\Theta(n)$	$\Theta(n)$	$O(n \log n)$	$O(n)$	$O(n \log n)$
Insert	$\Theta(n)$	$\Theta(i)$	$O(\log n)$	$\Theta(1)$	$O(\log n)$
Find	$O(n)$	$O(n)$	$O(\log n)$	$\Theta(1)$	$O(\log n)$
Membership	$O(n)$	$O(n)$	$O(\log n)$	$\Theta(1)$	$O(\log n)$
Remove	$\Theta(n)$	$\Theta(i)$	$O(\log n)$	$\Theta(1)$	$O(\log n)$
Sort	$O(n \log n)$	$O(n \log n)$	0	N/A	0
Index	$\Theta(1)$	$\Theta(i)$	$O(\log n)$	$\Theta(1)$	$O(\log n)$

3.7   CHOOSING A DATA STRUCTURE

As we have seen, the complexity of operations over data structures can be instrumental in choosing the appropriate data structure for a task. Consequently, it is beneficial to compare the asymptotic complexities of common operations over various data structures. Table 3.2 shows several commonly-used functions. Table 3.3 gives the asymptotic complexities of the algorithms used by these functions, for data structures containing n elements.

Arrays are ideal for computations requiring very fast random access. Lists are ideal for accumulating sequences of unknown lengths. The immutable set implementation in the Set module provides very fast set-theoretic operations union, intersection and difference. Hash tables are ideal for random access via keys that are not consecutive integers, e.g. sparsely-distributed keys or non-trivial types like strings.

The F# standard library contains other data structures such as mutable sets based upon hash tables. These do not offer fast set-theoretic operations but testing for membership is likely to be several times faster with mutable rather than immutable sets.

Having examined the various containers built-in to F#, we shall now examine a generalisation of these containers which is easily handled in F#.

3.8 SEQUENCES

The .NET collection called IEnumerable is known as Seq in F#. This collection provides an abstract representation of any lazily evaluated sequences of values. A lazily evaluated sequence only evaluates the elements when they are required, i.e. the sequence does not need to be stored explicitly.

The list, array, set, map and hash table containers all implement the Seq interface and, consequently, can be used as abstract sequences.

For example, the following function doubles each value in a sequence:

```
> let double s =
    Seq.map (( * ) 2) s;;
val double : #seq<int> -> seq<int>
```

Note that the argument to the double function is inferred to have the type #seq<int>, denoting objects of any class that implements the Seq interface, but the result of double is of the type seq<int>.

For example, this double function can be applied to arrays and sets:

```
> double [|1; 2; 3|];;
val it : seq<int> = [2; 4; 6]

> double (set [1; 2; 3]);;
val it : seq<int> = [2; 4; 6]
```

Sequences can be generated from functions using the Seq.generate function. In particular, such sequences can be infinitely long. The generate function accepts three functions as arguments. The first argument opens the stream. The second argument generates an option value for each element with None denoting the end of the stream. The third argument closes the stream.

This abstraction has various uses:

- Functions that act sequentially over data structures can be written more uniformly in terms of Seq. For example, by using Seq.fold instead of List.fold_left or Array.fold_left.

- Simplifies code interfacing different types of container.

- The higher-order Seq.map and Seq.filter functions are lazy, so a series of maps and filters is more space efficient and is likely to be faster using sequences rather than lists or arrays.

Sequences are useful in many situations but their main benefit is simplicity.

3.9 HETEROGENEOUS CONTAINERS

The array, list, set, hash table and map containers are all *homogeneous* containers, i.e. for any such container, the elements must all be of the same type. Containers of elements which may be of one of several different types can also be useful. These are known as *heterogeneous* containers.

Heterogeneous containers can be defined in F# by first creating a variant type which combines the types allowed in the container. For example, values of this variant type number may contain data representing elements from the sets $\mathbb{Z}$, $\mathbb{R}$ or $\mathbb{C}$:

```
> type number =
    | Integer of int
    | Real of float
    | Complex of float * float;;
```

A homogeneous container over this unified type may then be used to implement a heterogeneous container. The elements of a container of type number list can contain Integer, Real or Complex values:

```
> let nums = [Integer 1; Real 2.0; Complex (3.0, 4.0)];;
val nums : number list
```

Let us consider a simple function to act upon the effectively heterogeneous container type number list. A function to convert values of type number to the built-in type Math.Complex may be written:

```
> let complex_of_number = function
    | Integer n -> complex (float n) 0.0
    | Real x -> complex x 0.0
    | Complex(x, y) -> complex x y;;
val complex_of_number : number -> Math.Complex
```

For example, mapping the complex_of_number function over the number list called nums gives a Complex.t list:

```
> List.map complex_of_number nums;;
val it : Complex.t list =
  [1.0r + 0.0i; 2.0r + 0.0i; 3.0r + 4.0i]
```

The list, set, hash table and map data structures can clearly be useful in a scientific context, in addition to conventional arrays and the heterogeneous counterparts of these data structures. However, a major advantage of F# lies in its ability to create and manipulate new, custom-made data structures. We shall now examine this aspect of scientific programming in F#.

3.10 TREES

In addition to the built-in data structures, the ease with which the F# language allows tuples, records and variant types to be handled makes it an ideal language for creating and using new data structures. Trees are the most common such data structure.

A tree is a self-similar data structure used to store data hierarchically. The origin of a tree is, therefore, itself a tree, known as the *root* node. As a self-similar, or

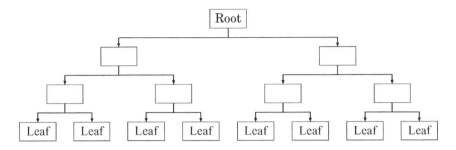

Figure 3.10   A perfectly-balanced binary tree of depth $x = 3$ containing $2^{x+1} - 1 = 15$ nodes, including the *root* node and $2^x = 8$ *leaf* nodes.

recursive, data structure, every node in a tree may contain further trees. A root that contains no further trees marks the end of a lineage in the tree and is known as a *leaf* node.

The simplest form of tree is a recursive data structure containing an arbitrarily-long list of trees. This is known as an n-ary tree and may be represented in F# by the type:

```
> type tree = Node of tree list;;
```

A perfectly balanced binary tree of depth d is represented by an empty node for $d = 0$ and a node containing two balanced binary trees, each of depth $d - 1$, for $d > 0$. This simple recurrence relation is most easily implemented as a purely functional, recursive function:

```
> let rec balanced_tree = function
  | 0 -> Node []
  | n ->
     Node [balanced_tree(n-1); balanced_tree(n-1)];;
val balanced_tree : int -> tree
```

The tree depicted in figure 3.10 may then be constructed using:

```
> let example = balanced_tree 3;;
val example : tree =
  Node
    [Node [Node [Node []; Node []];
           Node [Node []; Node []]];
     Node [Node [Node []; Node []];
           Node [Node []; Node []]]]
```

We shall use this `example` tree to demonstrate more sophisticated computations over trees.

Functions over the type `tree` are easily written. For example, the following function counts the number of leaf nodes:

```
> let rec leaf_count = function
```

```
  | Node [] -> 1
  | Node list ->
      Seq.fold (fun s t -> s + leaf_count t) 0 list;;
val leaf_count : tree -> int
> leaf_count example;;
val it : int = 8
```

Trees represented by the type tree are of limited utility as they cannot contain additional data in their nodes. An equivalent tree which allows arbitrary, polymorphic data to be placed in each node may be represented by the type:

```
> type 'a ptree = PNode of 'a * 'a ptree list;;
```

As a trivial example, the following function traverses a value of type tree to create an equivalent value of type ptree which contains a zero in each node:

```
> let rec zero_ptree_of_tree (Node list) =
    PNode(0, List.map zero_ptree_of_tree list);;
val zero_ptree_of_tree : tree -> int ptree
```

For example:

```
> zero_ptree_of_tree (Node [Node []; Node []]);;
val it : int ptree =
  PNode (0, [PNode (0, []); PNode (0, [])])
```

As a slightly more interesting example, consider a function to convert a value of type tree to a value of type ptree, storing unique integers in each node of the resulting tree.

We begin by defining a higher-order function aux1 that accepts a function f that will convert a tree into a ptree and uses this to convert the head of a list h and prepend it onto a tail list t in an accumulator that also contains the current integer n (i.e. the accumulator will be of type int * int ptree list):

```
> let aux1 f h (n, t) =
    let n, h = f h n
    n, h :: t;;
val aux1 :
  ('a -> 'b -> 'c * 'd) -> 'a -> 'b * 'd list ->
    'c * 'd list
```

A function aux2 to convert a tree into a ptree can be written by folding the aux1 function over the counter n and child trees list, accumulating a new counter n' and a list plist of child trees.

```
> let rec aux2 (Node list) n =
    let n', plist =
      List.fold_right (aux1 aux2) list (n + 1, [])
    n', PNode(n, plist);;
val aux2 : tree -> int -> int * int ptree
```

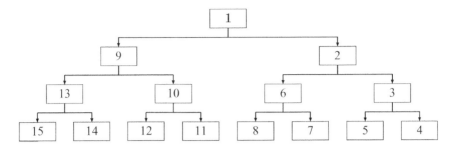

Figure 3.11   The result of inserting an integer counter into each node of the tree depicted in figure 3.10 using the `counted_ptree_of_tree` function.

For example:

```
> aux2 (Node[Node []; Node []; Node[]]) 1;;
val it : int * int ptree =
  (5, PNode (1, [PNode (4, []);
                 PNode (3, []);
                 PNode (2, [])]))
```

The `aux2` function can be used to convert a `tree` into a `ptree` with counted nodes by supplying an initial counter of 1 and stripping off the final value of the counter from the return value of `aux2` using the built-in function `snd` to extract the second part of a 2-tuple:

```
> let rec counted_ptree_of_tree tree =
    snd (aux2 tree 1);;
val counted_ptree_of_tree : tree -> int ptree
```

Applying this function to our example tree produces a more interesting result (illustrated in figure 3.11):

```
> counted_ptree_of_tree example;;
val it : int ptree =
  PNode (1,
    [PNode (9, [PNode (13, [PNode (15, []);
                            PNode (14, [])]);
                PNode (10, [PNode (12, []);
                            PNode (11, [])])]);
     PNode (2, [PNode (6, [PNode (8, []);
                           PNode (7, [])]);
                PNode (3, [PNode (5, []);
                           PNode (4, [])])])])
```

Note that the use of a right fold in the `aux2` function allowed the child trees to be accumulated into a list in the same order that they were given but resulted in numbering from right to left.

In practice, storing the maximum depth remaining in each branch of a tree can be useful when writing functions to handle trees. Values of our generic tree type may be converted into the ptree type, storing the integer depth in each node, using the following function:

```
> let rec depth_ptree_of_tree (Node list) =
    let list = List.map depth_ptree_of_tree list
    let depth_of (PNode(d, _)) = d
    let depth = Seq.fold (+) (-1) (Seq.map depth_of list)
    PNode(depth + 1, list);;
val depth_ptree_of_tree : tree -> int ptree
```

This function first maps itself over the child trees to get their depths. Then it folds an anonymous function that computes the maximum of each depth (note the use of partial application of max) with an accumulator that is initially -1. Finally, a node of a ptree is constructed from the depth and the converted children.

Applying this function to our example tree produces a rather uninteresting, symmetric set of branch depths:

```
> depth_ptree_of_tree example;;
val it : int ptree =
  PNode (3,
    [PNode (2,
      [PNode (1, [PNode (0, []); PNode (0, [])]);
       PNode (1, [PNode (0, []); PNode (0, [])])]);
     PNode (2,
      [PNode (1, [PNode (0, []); PNode (0, [])]);
       PNode (1, [PNode (0, []); PNode (0, [])])])])
```

Using a tree of varying depth provides a more interesting result. The following function creates an unbalanced binary tree, effectively representing a list of increasingly-deep balanced binary trees in the left children:

```
> let rec unbalanced_tree n =
    let aux n t = Node [balanced_tree n; t]
    List.fold_right aux [0 .. n - 1] (Node []);;
val unbalanced_tree : int -> tree
```

This can be used to create a wonky tree:

```
> let wonky = unbalanced_tree 3;;
val wonky : tree =
  Node
    [Node [];
     Node
      [Node [Node []; Node []];
       Node [Node [Node [Node []; Node []];
                   Node [Node []; Node []]];
             Node []]]]
```

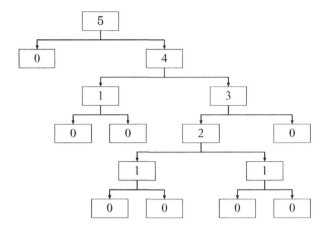

Figure 3.12 An unbalanced binary tree with the remaining depth stored in every node.

Converting this wonky tree into a tree containing the remaining depth in each node we obtain a more interested result (illustrated in figure 3.12):

```
> depth_ptree_of_tree wonky;;
val it : int ptree =
  PNode (5,
   [PNode (0, []);
    PNode (4,
      [PNode (1, [PNode (0, []); PNode (0, [])]);
       PNode (3,
        [PNode (2,
          [PNode (1, [PNode (0, []); PNode (0, [])]);
           PNode (1, [PNode (0, []); PNode (0, [])])]);
         PNode (0, [])])])])])
```

However, the ability to express a tree in which each node may have an arbitrary number of branches often turns out to be a hindrance rather than a benefit. Consequently, the number of branches allowed at each node in a tree is often restricted to one of two values:

- zero branches for leaf nodes and

- a constant number of branches for all other nodes.

Trees which allow only zero or two branches, known as *binary trees*, are particularly prolific as they form the simplest class of such trees[14], simplifying the derivations of the complexities of operations over this kind of tree.

---

[14]If only zero or one "branches" are allowed at each node then the tree is actually a list (see section 3.8).

Although binary trees could be represented using the `tree` or `ptree` data structures, this would require the programmer to ensure that all functions acting upon these types produced node lists containing either zero or two elements. In practice, this is likely to become a considerable source of human error and, therefore, of programmer frustration. Fortunately, when writing in F#, the type system can be used to enforce the use of a valid number of branches at each node. Such automated checking not only removes the need for careful inspection by the programmer but also removes the need to perform run-time checks of the data, improving performance. A binary tree analogous to our `tree` data structure can be defined as:

```
> type bin_tree =
    | Leaf
    | Node of bin_tree * bin_tree;;
```

A binary tree analogous to our `ptree` data structure can be defined as:

```
> type 'a pbin_tree =
    | Leaf of 'a
    | Node of 'a pbin_tree * 'a * 'a pbin_tree;;
```

Values of type `ptree` which represent binary trees may be converted to this `pbin_tree` type using the following function:

```
> let rec pbin_tree_of_ptree = function
    | PNode(v, []) -> Leaf v
    | PNode(v, [l; r]) ->
        Node(pbin_tree_of_ptree l,
             v,
             pbin_tree_of_ptree r)
    | _ -> invalid_arg "pbin_tree_of_ptree";;
val pbin_tree_of_ptree : 'a ptree -> 'a pbin_tree
```

For example, the arbitrary-branching-factor `example` tree may be converted into a binary tree using the `pbin_tree_of_ptree` function:

```
> pbin_tree_of_ptree (depth_ptree_of_tree example);;
val it : int pbin_tree =
  Node
    (Node
      (Node (Leaf 0, 1, Leaf 0),
       2,
       Node (Leaf 0, 1, Leaf 0)),
     3,
     Node
      (Node (Leaf 0, 1, Leaf 0),
       2,
       Node (Leaf 0, 1, Leaf 0)))
```

Note that the `'a pbin_tree` type, which allows arbitrary data of type `'a` to be stored in all nodes, could be usefully altered to an `('a, 'b) t` type which allows

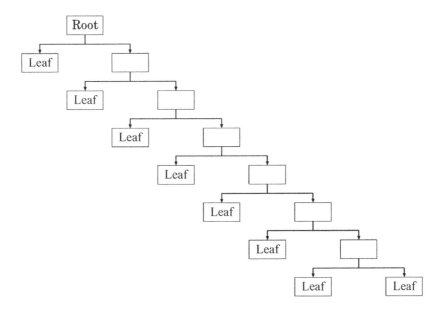

Figure 3.13   A maximally-unbalanced binary tree of depth $x = 7$ containing $2x + 1 = 15$ nodes, including the *root* node and $x + 1 = 8$ *leaf* nodes.

arbitrary data of type `'a` to be stored in leaf nodes and arbitrary data of type `'b` to be stored in all other nodes:

```
> type ('a, 'b) t =
    | Leaf of 'a
    | Node of 'b * ('a, 'b) t * ('a, 'b) t;;
```

Having examined the fundamentals of tree-based data structures, we shall now examine the two main categories of trees: balanced trees and unbalanced trees.

3.10.1 Balanced trees

Balanced trees, in particular balanced binary trees, are prolific in computer science literature. As the simplest form of tree, binary trees simplify the derivation of algorithmic complexities. These complexities often depend upon the depth of the tree. Consequently, in the quest for efficient algorithms, data structures designed to maintain approximately uniform depth, known as *balanced trees*, are used as the foundation for a wide variety of algorithms.

A balanced tree (as illustrated in figure 3.10) can be defined as a tree for which the difference between the minimum and maximum depths tends to a finite value for any such tree containing n nodes in the limit[15] $n \to \infty$. Practically, this condition is

---

[15]Although taking limits over integer-valued variables may seem dubious, the required proofs can, in fact, be made rigorous.

often further constricted to be that the difference between the minimum and maximum depths is no more than a small constant.

The efficiency of balanced trees stems from their structure. In terms of the number of nodes traversed, any node in a tree containing either n nodes or n leaf nodes may be reached by $O(\log n)$ traversals from the root.

For example, the set and map data structures provided in the F# standard library both make use of balanced binary trees internally. This allows them to provide single-element insertion, removal and searching in $O(\ln n)$ time-complexity.

For detailed descriptions of balanced tree implementation, we refer the eager reader to the relevant computer science literature [3] and, in particular, to an article in The F#.NET Journal [12]. However, although computer science exploits balanced trees for the efficient asymptotic algorithmic complexities they provide for common operations, which is underpinned by their balanced structure, the natural sciences can also benefit from the use of *unbalanced* trees.

3.10.2   Unbalanced trees

Many forms of data commonly used in scientific computing can be usefully represented hierarchically, in tree data structures. In particular, trees which store exact information in leaf nodes and approximate information in non-leaf nodes can be of great utility when writing algorithms designed to compute approximate quantities. In this section, we shall consider the development of efficient functions required to simulate the dynamics of particle systems, providing implementations for one-dimensional systems of gravitating particles. We begin by describing a simple approach based upon the use of a flat data structure (an array of particles) before progressing on to a vastly more efficient, hierarchical approach which makes use of approximate methods and the representation of particle systems as unbalanced binary trees. Finally, we shall discuss the generalisation of the unbalanced-tree-based approach to higher dimensionalities and different problems.

In the context of a one-dimensional system of gravitating particles, the mass $m > 0 \in \mathbb{R}$ and position $r \in \mathbb{R}^1$ of a particle may be represented by the record:

```
> type particle = { m: float; r: float };;
```

A function `force2` to compute the gravitational force (up to a constant coefficient):

$$F = \frac{m_1 m_2}{|\mathbf{r}_2 - \mathbf{r}_1|^3} (\mathbf{r}_2 - \mathbf{r}_1)$$

between two particles, `p1` and `p2`, may then be written:

```
> let force2 p1 p2 =
    let d = p2.r - p1.r
    p1.m * p2.m / (d * abs_float d);;
val force2 : particle -> particle -> float
```

For example, the force on a particle p_1 of mass $m_1 = 1$ at position $\mathbf{r}_1 = 0.1$ due to a particle p_2 of mass $m_2 = 3$ at position $r_2 = 0.8$ is:

$$F = \frac{1 \times 3}{(0.8 - 0.1)^2} = \frac{300}{49} \simeq 6.12245$$

```
> force2 { m = 1.0; r = 0.1 } { m = 3.0; r = 0.8 };;
val it : float = 6.12244897959183554
```

The `particle` type and `force2` function underpin both the array-based and tree-based approaches outlined in the remainder of this section.

3.10.2.1   *Array-based force computation*   The simplest approach to computing the force on one particle due to a collection of other particles is to store the other particles as a particle array and simply loop through the array, accumulating the result of applying the `force2` function. This can be achieved using a fold:

```
> let array_force p ps =
    Array.fold_left (fun f p2 -> f + force2 p p2)
      0.0 ps;;
val array_force : particle -> particle array -> float
```

This function can be demonstrated on randomised particles. A particle with random mass $m \in [0 \ldots 1)$ and position $r \in [0 \ldots 1)$ can be created using the function:

```
> let rand = new System.Random();;
val rand : System.Random
> let random_particle _ =
    { m = rand.NextDouble(); r = rand.NextDouble() };;
val random_particle : 'a -> particle
```

A random array of particles can then be created using the function:

```
> let random_array n = Array.init n random_particle;;
val random_array : int -> particle array
```

The following computes the force on a random particle `origin` due to a random array of 10^5 particles `system`:

```
> let origin = random_particle ();;
val origin : particle
> let system = random_array 100000;;
val system : particle array
> array_force origin sys;;
val it : float = 2362356950.0
```

Computing the force on each particle in a system of particles is the most fundamental task when simulating particle dynamics. Typically, the whole system is simulated in discrete time steps, the force computed for each particle being used to calculate the velocity and acceleration of the particle in the next time step. In a system of n particles, the `array_force` function applies the `force2` function exactly $n - 1$ times. Thus, using the `array_force` function to compute the force on all n particles would require $\Theta(n^2)$ time-complexity. This quadratic complexity forms the bottleneck of the whole simulation. Hence, the `array_force` function is an ideal target for optimization.

In this case, the array-based function to compute the force on a particle takes around a second. Applying this function to each of the 10^5 particles would, therefore, be expected to take almost a day. Thus, computing the update to the particle dynamics for a single time step is likely to take at least a day. This is highly undesirable. Moreover, there is no known approach to computing the force on a particle which both improves upon the $\Theta(n^2)$ asymptotic complexity whilst also retaining the apparent exactness of the simple, array-based computation we have just outlined.

In computer science, algorithms are optimized by carefully designing alternative algorithms which possess better complexities whilst also producing *exactly* the same results. This pedantry concerning accuracy is almost always appropriate in computer science. However, many subjects, including the natural sciences, can benefit enormously from relinquishing this exactness in favour of artful approximation. In particular, the computation of approximations known to be accurate to within a quantified error. As we shall now see, the performance of the array-based function to compute the force on a particle can be greatly improved upon by using an algorithm designed to compute an approximation to the exact result.

Promoting the adoption of approximate techniques in scientific computations can be somewhat of an uphill struggle. Thus, we shall now devote a little space to the arguments involved.

Often, when encouraged to convert to the use of approximate computations, many scientists respond by wincing and citing an article concerning the weather and the wings of a butterfly. Their point is, quite validly, that the physical systems most commonly simulated on computer are chaotic. Indeed, if the evolution of such a system could be calculated by reducing the physical properties to a solvable problem, there would be no need to simulate the system computationally.

The chaotic nature of simulated systems raises the concern that the use of approximate methods might change the simulation result in an unpredictable way. This is a valid concern. However, virtually all such simulation methods are already inherently approximate. One approximation is made by the choice of simulation procedure, such as the Verlet method for numerically integrating particle dynamics over time [8]. Another approximation is made by the use of finite-precision arithmetic (discussed in chapter 4). Consequently, the results of simulations should never be examined at the microscopic level but, rather, via quantities averaged over the whole system. Thus, the use of approximate techniques does not worsen the situation.

We shall now develop approximation techniques of controllable accuracy for computing the force of a particle due to a given system of particles, culminating in the implementation of a `force` function which provides a substantially more efficient alternative to the `array_force` function for very reasonable accuracies.

3.10.2.2 Tree-based force computation

In general, the strength of particle-particle interactions diminishes with distance. Consequently, the force exerted by a collection of distant particles may be well-approximated by grouping the collection into a pseudo-particle. In the case of gravitational interactions, this corresponds to grouping the effects of large numbers of smaller masses into small numbers of larger masses. This grouping effect can be obtained by storing the particle system in a tree data structure in which branches of the tree represent spatial subdivision, leaf nodes store exact particle information and other nodes store the information required to make approximations pertaining to the particles in the region of space represented by their lineage of the tree.

The spatial partitioning of a system of particles at positions r_i may be represented by an unbalanced binary tree of the type:

```
> type partition =
    | Leaf of particle list
    | Node of partition * particle * partition;;
```

Leaf nodes in such a tree contain a list of particles at the same or at similar positions. Other nodes in the tree contain left and right branches (which will be used to represent implicit subranges $[x_0, x_1)$ and $[x_1, x_2)$ of the range $[x_0, x_2)$ where $x_1 = \frac{1}{2}(x_0 + x_2)$, respectively) and the mass and position of a pseudo-particle chosen to approximate the summed effects of all particles farther down the tree.

The mass m_p and position r_p of a pseudo-particle approximating the effects of a set of particles (m_i, r_i) is given by the sum of the masses and the weighted average of the positions of particles in the child branches, respectively:

$$m_p = \sum_i m_i \qquad r_p = \frac{1}{m_p} \sum_i m_i r_i$$

In order to clarify the accumulator that contains these two sums and the current sub-partition, we shall use a record type called `accu`:

```
> type accu = { mp: float; mprp: float; p: partition };;
```

A function to compute the accumulator for a `Leaf` partition from a list of particles `ps` may be written:

```
> let make_leaf ps =
    let sum = List.fold_left ( + ) 0.0
    let m = List.map (fun p -> p.m) ps |> sum
    let mr = List.map (fun p -> p.m * p.r) ps |> sum
    {mp = m; mprp = mr; p = Leaf ps};;
val make_leaf : particle list -> accu
```

For example:

```
> make_leaf [{m = 1.0; r = -1.0}; {m = 3.0; r = 1.0}];;
val it : accu =
  {mp = 4.0;
```

```
mprp = 2.0;
p = Leaf [{m = 1.0; r = -1.0;};
          {m = 3.0; r = 1.0;}];}
```

A function to compose two accumulators into a Node partition that includes a pseudo-particle approximant may be written:

```
> let make_node a1 a2 =
    let mp = a1.mp + a2.mp
    let mprp = a1.mprp + a2.mprp
    {mp = mp;
     mprp = mprp;
     p = Node(a1.p, {m = mp; r = mprp / mp}, a2.p)};;
val make_node : accu -> accu -> accu
```

A particle system consists of the lower and upper bounds of the partition and the partition itself:

```
> type system =
    { lower: float; tree: partition; upper: float };;
```

The partition must be populated from a flat list of particles. This can be done by recursively bisecting the range of the partition until either it contains 0 or 1 particles or until the range is very small:

```
> let rec partition x0 ps x2 =
    match ps with
    | [] | [_] as ps -> make_leaf ps
    | ps when x2 - x0 < epsilon_float -> make_leaf ps
    | ps ->
        let x1 = (x0 + x2) / 2.0
        let ps1, ps2 =
          List.partition (fun p -> p.r < x1) ps
        make_node
          (partition x0 ps1 x1)
          (partition x1 ps2 x2);;
val partition : float -> particle list -> float -> accu
```

The List.partition function is used to divide the list of particles ps into those lying in the lower half of the range (ps1) and those lying in the upper half (ps2).

This partition function can be used to create a system from an array of particles:

```
> let rec make_system lower ps upper =
    {lower = lower;
     tree = (partition lower (List.of_seq ps) upper).p;
     upper = upper};;
val make_system :
  float -> particle array -> float -> system
```

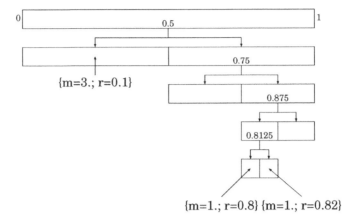

Figure 3.14 An unbalanced binary tree used to partition the space $r \in [0, 1)$ in order to approximate the gravitational effect of a cluster of particles in a system.

For example, consider the partition that results from 3 particles:

```
> let particles =
    [|{m = 3.0; r = 0.1}; {m = 1.0; r = 0.8};
     {m = 1.0; r = 0.82}|];;
val particles : particle array
> make_system 0.0 particles 1.0;;
val it : system =
  {lower = 0.0;
   tree = Node (Leaf [{m = 3.0; r = 0.1;}],
           {m = 5.0; r = 0.384;},
            Node (Leaf null, {m = 2.0; r = 0.81;},
             Node
              (Node (Leaf [{m = 1.0; r = 0.8;}],
                {m = 2.0; r = 0.81;},
                Leaf [{m = 1.0; r = 0.82;}]),
               {m = 2.0; r = 0.81;}, Leaf null)));
   upper = 1.0;}
```

This tree is illustrated in figure 3.14. Note that the pseudo-particle at the root node of the tree correctly indicates that the total mass of the system is $m = 3 + 1 + 1 = 5$ and the centre of mass is at $r = \frac{1}{5}(3 \times 0.1 + 0.8 + 0.82) = 0.384$.

We shall now consider the force on the particle at $r = 0.1$, exerted by the other particles at $r = 0.8$ and 0.82. The force can be calculated exactly, in arbitrary units, as:

$$F = \sum_j \frac{m_i m_j}{(r_j - r_i)^2} = \frac{3 \times 1}{0.7^2} + \frac{3 \times 1}{0.72^2} \simeq 11.9095$$

In this case, the force on the particle at $r_i = 0.1$ can also be well-approximated by grouping the effect of the other two particles into that of a pseudo-particle. From the tree, the pseudo-particle for the range $\frac{1}{2} \leq r_p < 1$ is $\{\texttt{m = 2.0; r = 0.81}\}$. Thus, the force may be well approximated by:

$$F \simeq \frac{m_p m_i}{(r_p - r_i)^2} = \frac{3 \times 2}{0.71^2} \simeq 11.9024$$

where m_p and r_p are the mass and centre of mass of the pseudo-particle, respectively.

Given the representation of a particle system as an unbalanced partition tree, the force on any given "origin" particle due to the particles in the system can be computed very efficiently by recursively traversing the tree either until a pseudo-particle in a non-leaf node is found to approximate the effects of the particles in its branch of the tree to sufficient accuracy or until real particles are found in a leaf node. This approach can be made more rigorous by bounding the error of the approximation.

The simplest upper bound of error is obtained by computing the difference between the minimum and maximum forces which can be obtained by a particle distribution satisfying the constraint that it must produce the pseudo-particle with the appropriate mass and position. If $r_i \notin [x_0, x_2)$, the force F is bounded by the force due to masses at either end of the range and the force due to all the mass at the centre of mass:

$$\frac{m_p m_i}{(r_p - r_i)^2} \leq F \leq \frac{m_p m}{x_2 - x_0} \left(\frac{c - x_0}{(x_2 - r)^2} + \frac{x_2 - c}{(x_0 - r)^2} \right)$$

For example, the bounds of the force in the previous example are given by $r = 0.1$, $m = 3$, $x_0 = 0.5$, $c = 0.81$, $x_2 = 1$ and $M = 2$:

$$\frac{3 \times 2}{(0.81 - 0.1)^2} \leq F \leq \frac{3 \times 2}{1 - 0.5} \left(\frac{0.81 - 0.5}{(1 - 0.1)^2} + \frac{1 - 0.81}{(0.5 - 0.1)^2} \right)$$
$$11.9024 \leq F \leq 18.8426$$

If this error was considered to be too large, the function approximating the total force would bisect the range and sum the contributions from each half recursively. In this case, the lower half contains no particles and the upper half $[x_1, x_2) = [0.75, 1)$ is considered instead. This tightens the bound on the force to:

$$11.9024 \leq F \leq 12.5707$$

This recursive process can be repeated either until the bound on the force is tight enough or until an exact result is obtained.

The following function computes the difference between the upper and lower bounds of the force on an origin particle p due to a pseudo-particle pp representing a particle distribution in the spatial range from x0 to x2:

```
> let metric p pp x0 x2 =
    if x0 <= p.r && p.r < x2 then infinity else
    let sqr x = x * x
```

```
      let fmin = p.m * pp.m / sqr (p.r - pp.r)
      let g x y = (pp.r - x) / sqr(p.r - y)
      let fmax =
        p.m * pp.m / (x2 - x0) * (g x0 x2 - g x2 x0)
      fmax - fmin;;
val metric :
  particle -> particle -> float -> float -> float
```

Note that the `metric` function returns an infinite possible error if the particle p lies within the partition range $[x_0, x_2)$, as the partition might contain another particle at the same position ($\Rightarrow \frac{1}{0}$).

For example, these are the errors resulting from progressively finer approximations:

```
> metric { m = 3.0; r = 0.1 } { m = 2.0; r = 0.81 }
    0.0 1.0;;
val it : float = infinity
> metric { m = 3.0; r = 0.1 } { m = 2.0; r = 0.81 }
    0.5 1.0;;
val it : float = 6.94019227519524939
> metric { m = 3.0; r = 0.1 } { m = 2.0; r = 0.81 }
    0.75 1.0;;
val it : float = 0.668276868664461787
> metric { m = 3.0; r = 0.1 } { m = 2.0; r = 0.81 }
    0.75 0.875;;
val it : float = 0.277220270131675051
```

A function to compute an approximation to the total force on a particle p due to other particles in a system `sys` to within an error `delta` can be written:

```
> let rec force_aux x0 x2 = function
    | Leaf list ->
        Seq.fold (fun f p2 -> f + force2 p p2) 0.0 list
    | Node(left, pp, right) ->
        if metric p pp x0 x2 < delta then
          force2 p pp
        else
          let x1 = 0.5 * (x0 + x2)
          force_aux x0 x1 left + force_aux x1 x2 right;;
val force_aux : float -> float ->
> let force p system delta =
    force_aux system.lower system.upper sys.tree;;
val force : particle -> system -> float -> float
```

The tree representation of this particle system is easily constructed by folding our `insert` function over the array which was used to test the `array_force` function:

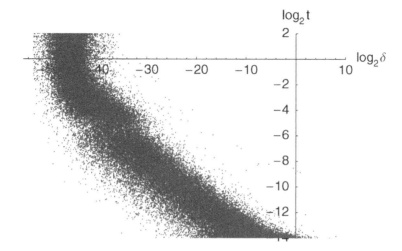

Figure 3.15   Measured performance of the tree-based approach relative to a simple array-based approach for the evaluation of long-range forces showing the resulting fractional error $\delta = |O - E|/E$ vs time taken $t = t_{\text{tree}}/t_{\text{array}}$ relative to the array-based method.

```
> let system = make_system 0.0 system 1.0;;
val system : system
```

The tree-based `force` function can compute controllably accurate approximations to the force on an origin particle due to a collection of other particles, trading accuracy for performance.

3.10.2.3 *Performance comparison*   Applying the `force` function with increasing permitted error results in a significant improvement in performance:

```
> force origin system 1e-9;;
val it : float = 2362356950.0
> force origin system 1e-6;;
val it : float = 2362356950.0
> force origin system 1e-3;;
val it : float = 2362356950.0
```

Moreover, the result does not change within 11 significant figures in this case!

From measurements of real-time performance (illustrated in figure 3.15), when requiring a force computation with an accuracy of one part in one million accuracy ($\log_2 \delta = -20$), the tree-based approach is approximately one thousand times faster ($\log_2 t \simeq -10$) than the array-based approach. Considering that, even when using the array-based approach, such computations are inherently approximate, a fractional error of 10^{-6} is a small price to pay for three orders of magnitude improvement in performance.

The tree-based approach we have just described is a simple form of what is now known as the *Fast Multipole Method* (**FMM**) [23]. Before being applicable to most physical systems, the approaches we have described must be generalized to higher dimensionalities. This generalisation is most easily performed by increasing the branching factor of the tree from 2 to 2^d for a d-dimensional problem. A more powerful generalisation involves associating the branches of the binary tree with subdivision along a particular dimension (either implicitly, typically by cycling through the dimensions, or explicitly, by storing the index of the subdivided dimension in the node of the tree). In particular, this allows anisotropic subdivision of space, i.e. some dimensions can be subdivided more than others. Anisotropic subdivision is useful in the context of anisotropic particle distributions, such as those found in many astrophysical simulations. One such method of anisotropic subdivision is known as the k-D tree.

3.10.3 Abstract syntax trees

Symbolic expressions can be represented very elegantly in F# as a form of tree known as an *abstract syntax tree*. This section describes a minimal implementation of symbolic expressions in F# and some simple functions to manipulate them.

3.10.3.1 Definition   Expressions composed of integers, variables, additions and multiplications may be represented by the variant type:

```
> type expr =
    | Int of int
    | Var of string
    | Add of expr * expr
    | Mul of expr * expr;;
```

For example, the following F# value represents the expression $(1 + x) \times 3$:

```
> Mul(Add(Int 1, Var "x"), Int 3);;
val it : expr = Mul(Add(Int 1, Var "x"), Int 3);;
```

Such expressions can be manipulated using functions over the `expr` type. Before we delve into this area, it is useful to provide a more convenient way to construct symbolic expressions as well as a more convenient way to visualize them in the F# interactive mode.

3.10.3.2 Easier construction   Symbolic expressions may be constructed by overloading the + and * operators. This is done by supplementing the definition of the variant type `expr` with static member functions (described in section 2.4.1.3):

```
> type expr =
    | Int of int
    | Var of string
    | Add of expr * expr
    | Mul of expr * expr
  with
    static member ( + ) (f, g) = Add(f, g)
```

```
static member ( - ) (f, g) =
  Add(f, Mul(Int(-1), g))
static member ( * ) (f, g) = Mul(f, g);;
```

The previous example may be constructed more clearly using these infix operators:

```
> (Int 1 + Var "x") * Int 3;;
val it : expr = Mul(Add(Int 1, Var "x"), Int 3)
```

In this case, the arithmetic operators naively compose subexpressions. More intelligent constructors can be used to maintain a more sophisticated and efficient representation of symbolic mathematical expressions.

3.10.3.3 *Evaluating expressions*

Perhaps the simplest function that acts upon a symbolic expression is one that evaluates the expression to get a value. Values of this `expr` type can be evaluated to the F# type `int` using the following `eval` function:

```
> let rec eval vars = function
    | Int n -> n
    | Var v -> List.assoc v vars
    | Add(f, g) -> eval vars f + eval vars g
    | Mul(f, g) -> eval vars f * eval vars g;;
val eval : (string * int) list -> expr -> int
```

The argument `vars` is an association list giving the values of variables that might appear in the expression being evaluated.

For example, the expression $(1 + x) \times 3$ may be evaluated in the context $x = 2$ to get $(1 + 2) \times 3 = 9$:

```
> eval ["x", 2] ((Int 1 + Var "x") * Int 3);;
val it : int = 9
```

By defining another variant type (typically called `value`) to represent the kinds of values that can result from evaluation, this approach can be generalized to evaluate more complicated expressions containing different kinds of numbers and even strings, compound types and functions. When expressions can define and apply arbitrary functions, this `eval` function becomes an interpreter for a simple programming language [14]. Many programming languages are implemented in this way and domain-specific languages have a variety of applications in scientific computing.

3.10.3.4 *Term rewriting*

Another important use of symbolic expressions is rewriting them to produce related expressions. This approach is often known as *term rewriting* and involves repeated applications of sets of rewrite rules designed to transform an expression into a different (but typically equivalent) form.

A higher-order `rewrite` function can be used to implement generic term rewriting:

```
> let rec rewrite rule expr =
```

```
    let expr' =
      match expr with
      | Int _ | Var _ as f -> rule f
      | Add(f, g) ->
          rule(rewrite rule f + rewrite rule g)
      | Mul(f, g) ->
          rule(rewrite rule f * rewrite rule g)
    if expr = expr' then expr else rewrite rule expr';;
val rewrite : (expr -> expr) -> expr -> expr
```

This `rewrite` function takes a function rule that transforms expressions and applies it to every subexpression of the given expression. For example, using `rewrite` to apply the identity rule to the expression $3 + x$ will apply the rule to the subexpressions 3, then x and then the whole expression $3 + x$. This `rewrite` function continues rewriting until the expression stops changing.

The rewrite rule $(f + g) \times h \to f \times h + g \times h$ can be used to expand products of sums and may be written:

```
> let rec expand = function
    | Mul(Add(f, g), h) | Mul(h, Add(f, g)) ->
        f * h + g * h
    | f -> f;;
val expand : expr -> expr
```

Note that F# does not know that `Mul(f, g)` and `Mul(g, f)` are equivalent, so we are careful to account for both alternatives in the pattern.

For example, the `rewrite` and `expand` functions may be used to expand the expression $(1 + x) \times 3$ to $1 \times 3 + x \times 3$:

```
> rewrite expand ((Int 1 + Var "x") * Int 3);;
val it : expr =
  Add(Mul(Int 1, Int 3), Mul(Var "x", Int 3))
```

Symbolic computations are of increasing importance in scientific computing and the ML family of languages are very powerful tools for such work, with existing applications ranging from accumulating contributions from billions of Feynman graphs in string theory [6] to the generation of the high-performance FFT implementations in software such as MATLAB [9].

We shall revisit the handling of symbolic expressions in F# later in this book, in the context of parsing (section 5.5.2), metaprogramming (section 9.12) and interoperating with computer algebra systems (section 11.3).

CHAPTER 4

NUMERICAL ANALYSIS

Computers can only perform finite computations. Consequently, computers only make use of finite precision representations of numbers. This has several important implications in the context of scientific computation.

This chapter provides an overview of the representations and properties of values of types int and float, used to represent members of the sets $\mathbb{Z}$ and $\mathbb{R}$, respectively. Practical examples demonstrating the robust use of floating-point arithmetic are then given. Finally, some other forms of arithmetic are discussed.

4.1 NUMBER REPRESENTATION

In this section, we shall introduce the representation of integer and floating-point numbers before outlining some properties of these representations.

4.1.1 Machine-precision integers

Positive integers are represented by several, least-significant binary digits (bits). For example, the number 1 is represented by the bits ...00001 and the number 11 is represented by the bits ...01011. Negative integers are represented in twos-

F# for Scientists. By Jon Harrop
Copyright © 2008 John Wiley & Sons, Inc.

Figure 4.1   Values i of the type int, called *machine-precision integers*, are an exact representation of a consecutive subset of the set of integers $i \in [l \ldots u] \subset \mathbb{Z}$ where l and u are given by min_int and max_int, respectively.

complement format. For example, the number -1 is represented by the bits ... 11111 and the number -11 is represented by the bits ... 10101.

Consequently, the representation of integers $n \in \mathbb{Z}$ by values of the type int is exact within a finite range of integers (illustrated in figure 4.1). This range is platform specific and may be obtained as the min_int and max_int values. On a 32-bit platform, the range of representable integers is substantial:

```
> min_int, max_int;;
val it : int * int = (-2147483648, 2147483647)
```

On a 64-bit platform, the range is even larger.

The binary representation of a value of type int may be obtained using the following function:

```
> let binary_of_int n =
    [ for i in Sys.word_size - 1 .. -1 .. 0 ->
        if (n >>> i) % 2 = 0 then "0" else "1" ]
    |> String.concat "";;
val binary_of_int : int -> string
```

For example, the 32-bit binary representations of 11 and -11 are:

```
> binary_of_int 11;;
val it : string = "00000000000000000000000000001011"
> binary_of_int (-11);;
val it : string = "11111111111111111111111111110101"
```

As we shall see in this chapter, the exactness of the int type can be used in many ways.

4.1.2   Machine-precision floating-point numbers

In science, many important numbers are written in *scientific notation*. For example, Avogadro's number is conventionally written $N_A = 6.02214 \times 10^{23}$. This notation essentially specifies the two most important quantities about such a number:

1. the most significant digits called the *mantissa*, in this case 6.02214, and

2. the offset of the decimal point called the *exponent*, in this case 23.

Figure 4.2 Values of the type float, called *double-precision floating-point numbers*, are an approximate representation of real-valued numbers, showing: a) full-precision (normalized) numbers (black), and b) denormalized numbers (gray).

Computers use a similar, finite representation called "floating point" which also contains a mantissa and exponent. In F#, floating-point numbers are represented by values of the type float. Roughly speaking, values of type int approximate real numbers between min_int and max_int with a constant absolute error of $\pm\frac{1}{2}$ whereas values of the type float have an approximately-constant relative error that is a tiny fraction of a percent.

In order to enter floating-point numbers succinctly, the F# language uses a standard "e" notation, equivalent to scientific number notation $a \times 10^b$. For example, the number 5.4×10^{12} may be represented by the value:

```
> 6.02214e23;;
val it : float = 6.02214e+23
```

As the name "floating point" implies, the use of a mantissa and an exponent allows the point to be "floated" to any of a wide range of offsets. Naturally, this format uses base-two (binary) rather than base-ten (decimal) and, hence, numbers are represented by the form $a \times 2^b$ where a is the mantissa and b is the exponent. Double-precision floating-point values consume 64-bits, of which 53 bits are attributed to the mantissa (including one bit for the sign of the number) and the remaining 11 bits to the exponent.

By default, F# interactive mode only displays a few digits of precision. In order to examine floating point numbers more accurately from the interactive mode, we shall replace the pretty printer used to visualize values of the float type with one that conveys more precision:

```
> fsi.AddPrinter(sprintf "%0.20g");;
val it : unit = ()
```

Compared to the type int, the exponent in a value of type float allows a huge range of real-valued numbers to be approximated. As for the type int, this range is given by the predefined values:

```
> min_float, max_float;;
val it : float * float =
  (2.22507385850720138e-308, 1.79769313486231571e+308)
```

Some useful values *not* in the set of real numbers $\mathbb{R}$ are also representable in floating-point number representation. Numbers out of range are expressed by the values -0.0, -Infinity ($-\infty$) and Infinity (∞). For example, in floating-point arithmetic $\frac{1}{0} = \infty$:

```
> 1.0 / 0.0;;
val it : float = Infinity
```

Floating point arithmetic includes a special value denoted nan in F# that is used to represent results that are "not a number". This is used when calculations do not return a real-valued number $x \in \mathbb{R}$, e.g. when a supplied parameter falls outside the domain of a function. For example[16], $\ln(-1) \notin \mathbb{R}$:

```
> log -1.0;;
val it : float = NaN
```

The domain of the log function is $0 \leq x \leq \infty$, with log 0.0 evaluating to -Infinity. and log infinity evaluating to Infinity.

The special value nan has some interesting and important properties. In particular, all comparisons involving nan return false except inequality of nan with itself:

```
> nan <= 3.0;;
val it : bool = false
> nan >= 3.0;;
val it : bool = false
> nan <> nan;;
val it : bool = true
```

However, although comparing nan with itself using equality indicates that nan is not equal to itself, the built-in compare function returns 0 when comparing nan with nan as a special case:

```
> compare nan nan;;
val it : int = 0
```

This ensures that collections built using the compare function, such as sets (described in section 3.4), do not leak when many nan values are inserted:

```
> set [nan; nan; nan];;
val it : Set<float> = seq [NaN]
```

In the case of $\ln(-1)$, the implementation of complex numbers provided in the Math.Complex module may be used to calculate the complex-valued result $\ln(-1) = \pi i$:

```
> log(complex -1.0 0.0);;
val it : Complex = 0.0r + 3.141592654i
```

Note that the built-in log function is overloaded to work transparently on the complex type as well as the float type. This makes it very easy to write functions that handle complex numbers in F#.

---

[16]Note that negative float literals do not need brackets, i.e. log -1.0 is equivalent to log(-1.0).

As well as min_float, max_float, infinity and nan, F# also predefines contains an epsilon_float value:

```
> epsilon_float;;
val it : float = 2.22044604925031308e-16
```

This is the smallest number that, when added to one, does not give one:

```
> 1.0 + epsilon_float;;
val it : float = 1.00000000000000022
```

Consequently, the epsilon_float value is seen in the context of numerical algorithms as it encodes the accuracy of the mantissa in the floating point representation. In particular, the square root of this number often appears as the accuracy of numerical approximants computed using linear approximations (leaving quadratics terms as the largest remaining source of error). This still leaves a substantially accurate result, suitable for most computations:

```
> 1.0 + sqrt epsilon_float;;
val it : float = 1.000000015
```

This was used in the definition of the d combinator in section 1.6.4.

The approximate nature of floating-point computations is often seen in simple calculations. For example, the evaluation of $\frac{1}{3}$ is only correct to 16 fractional digits:

```
> 1.0 / 3.0;;
val it : float = 0.333333333333333315
```

In particular, the binary representation of floating-point numbers renders many decimal fractions approximate. For example, although 1 is represented exactly by the type float, the decimal fraction 0.9 is not:

```
> 1.0 - 0.9;;
val it : float = 0.0999999999999999778
```

Many of the properties of conventional algebra over real-valued numbers can no longer be relied upon when floating-point numbers are used as a representation. For more details, see the relevant literature [17].

4.2 ALGEBRA

In real arithmetic, addition is associative:

$$(a + b) + c = a + (b + c)$$

In general, this is *not* true in floating-point arithmetic. For example, in floating-point arithmetic $(0.1 + 0.2) + 0.3 \neq 0.1 + (0.2 + 0.3)$:

```
> (0.1 + 0.2) + 0.3 = 0.1 + (0.2 + 0.3);;
val it : bool = false
```

In this case, round-off error results in slightly different approximations to the exact answer that are not exactly equal:

```
> (0.1 + 0.2) + 0.3, 0.1 + (0.2 + 0.3);;
val it : float * float =
  (0.60000000000000009, 0.59999999999999998)
```

Hence, even in seemingly simple calculations, values of type `float` should not be compared for exact equality.

More significant errors are obtained when dealing with the addition and subtraction of numbers with wildly different exponents. For example, in real arithmetic $1.3 + 10^{15} - 10^{15} = 1.3$ but in the case of `float` arithmetic:

```
> 1.3 + 1e15 - 1e15;;
val it : float = 1.25
```

The accuracy of this computation is limited by the accuracy of the largest magnitude numbers in the sum. In this case, these numbers are 10^{15} and -10^{15}, resulting in a significant error of 0.05 in this case.

In general, this problem boils down to the subtraction of similar-sized numbers. This includes adding a similar-sized number of the opposite sign.

The accuracy of calculations performed using floating-point arithmetic may often be improved by carefully rearranging expressions. Such rearrangements often result in more complicated expressions which are, therefore, slower to execute.

Consider the following function $f(x)$:

$$f_1(x) = \sqrt{1 + x} - 1$$

This involves the subtraction of a pair of similar numbers when $x \simeq 0$. This may be expressed in F# as:

```
> let f_1 x =
    sqrt (1.0 + x) - 1.0;;
val f_1 : float -> float
```

As expected, results of this function are significantly erroneous in the region $x \simeq 0$. For example:

$$f(10^{-15}) = \sqrt{\frac{1 + 10^{15}}{10^{15}}} - 1 \simeq 4.99999999999999875\ldots \times 10^{-16}$$

```
> f_1 1e-15;;
val it : float = 4.4408920985006262e-16
```

Note that the result given by F# differs from the mathematical result by ~ 0.056.

The f_1 function may be rearranged into a form which evades the subtraction of similar-sized numbers around $x \simeq 0$:

$$f_2(x) = \frac{x}{1 + \sqrt{1 + x}}$$

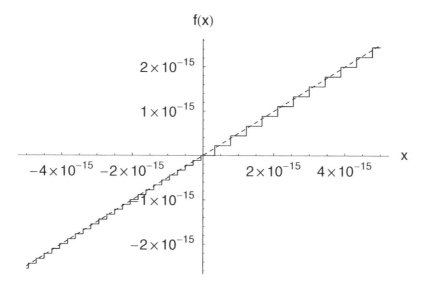

Figure 4.3   Accuracy of two equivalent expressions when evaluated using floating-point arithmetic: a) $f_1(x) = \sqrt{1+x} - 1$ (solid line), and b) $f_2(x) = x/(1 + \sqrt{1+x})$ (dashed line).

This alternative definition may be expressed in F# as:

```
> let f_2 x =
    x / (1.0 + sqrt (1.0 + x));;
val f_2 : float -> float
```

Although $f_1(x) = f_2(x) \ \forall \ x \in \mathbb{R}$, the f_2 form of the function is better behaved when evaluated in the region $x \simeq 0$ using floating-point arithmetic. For example, the value of the function at $x = 10^{-15}$ is much better approximated by f_2 than it was by f_1:

```
> f_2 1e-15;;
val it : float = 4.9999999999999994e-16
```

This is particularly clear on a graph of the two functions around $x \simeq 0$ (illustrated in figure 4.3).

4.3   INTERPOLATION

Due to the accumulation of round-off error, loops should not use loop variables of type float but, rather, use the type int and, if necessary, convert to the type float within the loop. Interpolation is an important example of this.

Comprehensions provide an easy way to demonstrate this. The following comprehension generates a list of values interpolating from 0.1 to 0.9 inclusive:

```
> [0.1 .. 0.2 .. 0.9];;
val it : float list = [0.1; 0.3; 0.5; 0.7; 0.9]
```

However, specifying the step size in this way is not numerically stable because repeated float additions lead to numerical error. For example, using a step size of 0.16 leads to the last value being omitted:

```
> [0.1 .. 0.16 .. 0.9];;
val it : float list = [0.1; 0.26; 0.42; 0.58; 0.74]
```

In this case, the result of repeatedly adding the approximate representation of d to that of x, starting with $x = l$, produced an approximation which was slightly greater than u. Thus, the sequence terminated too early. This produced a list containing five elements instead of the expected six.

As such functionality is commonly required in scientific computing, a robust alternative must be found.

Fortunately, this problem is easily solved by resorting to an exact form of arithmetic for the loop variable, typically `int` arithmetic, and converting to floating-point representation at a later stage. For example, the `interp` function may be written robustly by using an integer loop variable i:

$$x_i = x_1 + \frac{i-1}{n-1}(x_n - x_1)$$

for $i \in \{1 \ldots n\}$:

```
> let interp x1 xn n =
    [ for i in 1 .. n ->
        x1 + float(i - 1) * (xn - x1) / float(n - 1) ];;
val interp : float -> float -> int -> float list
```

Thanks to the use of an exact form of arithmetic, this function produces the desired result:

```
> interp 0.1 0.9 6;;
val it : float list = [0.1; 0.26; 0.42; 0.58; 0.74; 0.9]
```

In terms of comprehensions, this is equivalent to:

```
> [ for i in 0 .. 5 ->
      0.1 + float i * 0.8 / 5.0 ];;
val it : float list = [0.1; 0.26; 0.42; 0.58; 0.74; 0.9]
```

We shall now conclude this chapter with two simple examples of the inaccuracy of floating-point arithmetic.

4.4 QUADRATIC SOLUTIONS

The solutions of the quadratic equation $ax^2 + bx + c = 0$ are well known to be:

$$x_{1,2} = \frac{-b \pm \sqrt{b^2 - 4ac}}{2a}$$

The root $\sqrt{b^2 - 4ac}$ may be productively factored out of these expressions:

$$y = \sqrt{b^2 - 4ac}$$

$$x_1 = \frac{-b + y}{2a} \qquad x_2 = \frac{-b - y}{2a}$$

These values are easily calculated using floating-point arithmetic:

```
> let quadratic a b c =
    let y = sqrt (b * b - 4.0 * a * c)
    (-b + y) / (2.0 * a), (-b - y) / (2.0 * a);;
val quadratic : float -> float -> float -> float * float
```

However, when evaluated using floating-point arithmetic, these expressions can be problematic. Specifically, when $b^2 \gg 4ac$, subtracting $4ac$ from b^2 in the subexpression $b^2 - 4ac$ will produce an inaccurate result approximately equal to b^2. This results in $-b + \sqrt{b^2 - 4ac}$ becoming equivalent to $-b + b$ and, therefore, an answer of zero.

For example, using the conditions $a = 1$, $b = 10^9$ and $c = 1$, the correct solutions are $x \simeq -10^{-9}$ and -10^9 but the above implementation of the quadratic function rounds the smaller magnitude solution to zero:

```
> quadratic 1.0 1e9 1.0;;
val it : float * float = (0.0, -1e+9)
```

The accuracy of the smaller-magnitude solution is most easily improved by calculating the smaller-magnitude solution in terms of the larger-magnitude solution, as:

$$x_1 = \begin{cases} b \geq 0 & -\frac{y+b}{2a} \\ b < 0 & \frac{y-b}{2a} \end{cases} \qquad x_2 = \frac{c}{x_1 a}$$

This formulation, which avoids the subtraction of similar values, may be written:

```
> let quadratic a b c =
    let y = sqrt (b * b - 4.0 * a * c)
    let x1 = (if b < 0.0 then y-b else -y-b) / (2.0 * a)
    x1, c / (x1 * a);;
val quadratic : float -> float -> float -> float * float
```

This form of the quadratic function is numerically robust, producing a more accurate approximation for the previous example:

```
> quadratic 1.0 1e9 1.0;;
val it : float * float = (-1000000000.0, -1e-09)
```

Numerical robustness is required in a wide variety of algorithms. We shall now consider the evaluation of some simple quantities from statistics.

4.5 MEAN AND VARIANCE

In this section, we shall illustrate the importance of numerical stability using expressions for the mean and variance of a set of numbers.

The mean value $\bar{x}$ of a set of n numbers x_k is given by:

$$\bar{x} = \frac{1}{n} \sum_{k=1}^{n} x_k$$

The sum may be computed by folding addition over the elements and dividing by the length:

```
> let mean xs =
    Seq.fold ( + ) 0.0 xs / float(Seq.length xs);;
val mean : #seq<float> -> float
```

Note that this definition of the mean function can be applied to any sequence container (lists, arrays, sets etc.).

For example, the mean of $\{1, 3, 5, 7\}$ is $\frac{1}{4}(1 + 3 + 5 + 7) = 4$:

```
> mean [|1.0; 3.0; 5.0; 7.0|];;
val it : float = 4.0
```

Although the sum of a list of floating point numbers may be computed more accurately by accumulating numbers at different scales and then summing the result starting from the smallest scale numbers, the straightforward algorithm used by this mean function is often satisfactory. The same cannot be said of the straightforward computation of variance.

The variance σ^2 of a set x_k of numbers is typically written:

$$\sigma^2 = \frac{1}{n(n-1)} \left(\left(n \sum_{k=1}^{n} x_k^2 \right) - \left(\sum_{k=1}^{n} x_k \right)^2 \right)$$

Although variance is a strictly non-negative quantity, the subtraction of the sums in this expression for the variance may produce small, negative results when computed directly using floating-point arithmetic, due to the accumulation of rounding errors. This problem can be avoided by computing a numerically robust recurrence relation [17]:

$$M_1 = x_1 \qquad M_k = M_{k-1} + (x_k - M_{k-1})/k$$
$$S_1 = 0 \qquad S_k = S_{k-1} + (x_k - M_{k-1}) \times (x_k - M_k)$$

Thus, the variance may be computed more accurately using the following functions:

```
> let variance_aux (m, s, k) x =
    let m' = m + (x - m) / k
```

```
      m', s + (x - m) * (x - m'), k + 1.0;;
val variance_aux :
  float * float * float -> float ->
    float * float * float
> let variance xs =
    let _, s, n2 = Seq.fold aux (0.0, 0.0, 1.0) xs
    s / (n2 - 2.0);;
val variance : #seq<float> -> float
```

The auxiliary function `variance_aux` accumulates the recurrence relation when it is folded. The `variance` function simply folds this auxiliary function over the sequence to return a better-behaved approximation to the variance $\sigma^2 \simeq S_n/(n-1)$.

For example, the variance of $\{1, 3, 5, 7\}$ is $\sigma^2 = 6\frac{2}{3}$ and the `variance` function gives an accurate result:

```
> variance [|1.0; 3.0; 5.0; 7.0|];;
val it : float = 6.66666666666666696
```

The numerical stability of this `variance` function allows us to write a function to compute the standard deviation σ the obvious way, without having to worry about negative roots:

```
> let standard_deviation x =
    sqrt (variance x);;
val variance : #seq<float> -> float
```

Clearly numerically stable algorithms which use floating-point arithmetic can be useful. We shall now examine some other forms of arithmetic.

4.6 OTHER FORMS OF ARITHMETIC

As we have seen, the int and float types in F# represent numbers to a fixed, finite precision. Although computers can only perform arithmetic on finite-precision numbers, the precision allowed and used could be extended indefinitely. Such representations are known as arbitrary-precision numbers.

4.6.1 Arbitrary-precision integer arithmetic

The vast majority of F# programs use machine-precision integer arithmetic (described in section 4.1.1) because it is provided in hardware and, consequently, is extremely efficient. In some cases, programs are required to handle integers outside the range of machine-precision integers. In such cases, F# programs can use the built-in Math.BigInt type.

Arbitrary-precision integers may be specified as literals with an I suffix:

```
> open Math;;
```

```
> 123I;;
val it : bigint = 123I
```

The `BigInt` class provides some useful functions for manipulating arbitrary-precision integers such as a factorial function:

```
> BigInt.factorial 33I;;
val it : bignum = 8683317618811886495518194401280000000I
```

Arbitrary-precision integers are used in some parts of scientific computing, most notably in computer algebra systems.

4.6.2   Arbitrary-precision rational arithmetic

Rationals, fractions of the form $\frac{p}{q}$, $q > 0, p \in \mathbb{Z}$, may be represented exactly using *rational arithmetic*. This form of arithmetic uses arbitrary precision integers to represent p and q.

Compared to the type `float`, rational arithmetic allows arbitrary precision to be used for any value in $\mathbb{R}$. However, arithmetic operations on floating point numbers are $O(1)$ but arithmetic operations on arbitrary-precision rational numbers are asymptotically slower. The larger the numerators and denominators, the slower the calculations.

Arbitrary-precision rational arithmetic is implemented by the `BigNum` class. Rational literals are written `123N` or:

```
> 123N / 456N;;
val it : bignum = 41/152N
```

Note that the fraction is automatically reduced, in this case by dividing both numerator and denominator by 3.

For example, an arbitrary-precision factorial function may be written:

```
> let rec factorial = function
    | 0 -> 1N
    | n -> BigNum.of_int n * factorial (n - 1);;
val factorial : int -> bignum
> factorial 33;;
val it : bignum = 8683317618811886495518194401280000000N
```

Rational arithmetic, represented by the constructor `Ratio`, may then be used to calculate an approximation to:

$$e = \sum_{n=0}^{\infty} \frac{1}{n!}$$

```
> open Math;;
> let rec e = function
    | 0 -> 1N
    | n -> 1N / factorial n + e (n - 1);;
```

```
val e : int -> BigNum
```

For example, the first eighteen terms give the rational approximation to e of:

$$\sum_{n=0}^{17} \frac{1}{n!} = \frac{7437374403113}{2736057139200}$$

```
> e 17;;
val it : string = 7437374403113/2736057139200N
```

Converting this result from rational representation into a floating point number demonstrates that the result is very close to the value given by a native floating point evaluation of e:

```
> Math.BigNum.to_float(e 17);;
val it : float = 2.718281828
> exp 1.0;;
val it : float = 2.718281828
```

Rational arithmetic can be useful in many circumstances, including geometric computations.

4.6.3 Adaptive precision

Some problems can use ordinary precision floating-point arithmetic most of the time and resort to higher-precision arithmetic only when necessary. This leads to adaptive-precision arithmetic, which uses fast, ordinary arithmetic where possible and resorts to suitable higher-precision arithmetic only when required.

Geometric algorithms are an important class of such problems and a great deal of interesting work has been done on this subject [24]. This work is likely to be of relevance to scientists studying the geometrical properties of natural systems.

CHAPTER 5

INPUT AND OUTPUT

In this chapter, we examine the various ways in which an F# program can transfer data, including printing to the screen and saving and loading information on disc. In particular, we examine some sophisticated tools for F# which greatly simplify the task of designing and implementing programs to load files in well-defined formats.

5.1 PRINTING

The ability to print information on the screen or into a string or file is useful as a means of conveying the result of a program (if the result is simple enough) and providing run-time information on the current state of the program as well as providing extra information to aid debugging. Naturally, F# provides several functions to print to the screen. In particular, the `printf` function can be used to print a variety of different types and perform some simple formatting.

The `printf` function understands several *format specifiers*:

- `%s` print a `string`

- `%d` print an `int`

- `%f` print a `float`

- `%g` print a `float` in scientific notation

- `%a` print a value using a custom print function

- `%O` print a value of any type using the `ToString()` method.

- `%A` print a value of any type using the built-in structural pretty printer.

Special characters can also be printed:

- `\n` newline

- `\"` double-quotes

- `\t` tab

For example, the following prints a `string` in quotes, an `int` and then a `float`:

```
> printf "String: \"%s\", int: %d, float: %f\n" "foo" 3
    7.4;;
String: "foo", int: 3, float: 7.4
```

The following function uses the generic `%A` format specifier to print a value as a running F# interactive session would:

```
> printf "%A" (1, 2.3, "foo");;
(1, 2.3, "foo")
```

The ability to pretty print values of any type at run time is particularly useful during debugging. There are also related functions that use the same syntax to print into a string or to a channel.

5.1.1 Generating strings

Values can be printed into strings using the `sprintf` function. This function accepts the same format specifiers as the `printf` function.

For example, the following prints the same result as the previous example but returns a string containing the result:

```
> sprintf "String: \"%s\", int: %d, float: %f\n" "foo" 3
    7.4;;
val it : string = "String: \"foo\", int: 3, float: 7.4"
```

The `%a` format specifier is most easily elucidated using a example based upon the `sprintf` function. This format specifier expects two corresponding arguments, a print function `g` and its argument `x`:

```
> let f g x = sprintf "%a" g x;;
val f : (unit -> 'a -> string) -> 'a -> string
```

This is particularly useful when printing recursive data structures such as lists and trees as a print function can pass itself or another related function as an argument to `sprintf`.

The F# interactive mode can be supplemented with custom pretty printers for user-defined types by supplying a function to convert a value of that type into the corresponding string to the `fsi.AddPrinter` function. A function to convert a value of the type `expr` to a string may be written using the `sprintf` function and the `%d` and `%a` format specifiers:

```
> let rec string_of_expr () expr =
    | Int n -> sprintf "%d" n
    | Var v -> v
    | Add(f, g) ->
        sprintf "%a + %a"
          string_of_expr f string_of_expr g
    | Mul(f, g) ->
        sprintf "%a * %a"
          string_of_mul f string_of_mul g
  and string_of_mul () = function
    | Int _ | Mul _ as f -> string_of_expr () f
    | Add _ as f -> sprintf "(%a)" string_of_expr f;;
val string_of_expr : unit -> expr -> string
val string_of_mul : unit -> expr -> string
```

Printing of subexpressions inside a product are dispatched to the auxiliary function `string_of_mul` in order to bracket subexpressions that have a lower precedence. Specifically, any addition inside a multiplication must be bracketed when pretty printing an expression.

The previous example expression is now printed more comprehensibly by the F# interactive mode:

```
> (Int 1 + Var "x") * Int 3;;
val it : expr = (1 + x) * 3;;
```

Note that the subexpression $1 + x$ is bracketed correctly when it is printed because the `string_of_expr` function takes the relative precedence of multiplication over addition into account.

5.2 GENERIC PRINTING

In addition to printing values of specific types, the F# programming language also provides a `print_any` function (equivalent to `printf "%A"`) to print values of any type. For example, to print a list of 2-tuples in F# syntax:

```
> print_any [1, 2; 2, 3];;
[(1, 2); (2, 3)]
```

Analogously, there is an `any_to_string` function:

```
> any_to_string [1, 2; 2, 3];;
val it : string = "[(1, 2); (2, 3)]"
```

This is particularly useful when debugging, to avoid having to write conversion functions for each and every type.

5.3   READING FROM AND WRITING TO FILES

The act of saving data in a file is performed by opening the file, writing data to it and then closing the file. F# provides an elegant and efficient way to do this using functionality provided by .NET in the `System.IO` namespace combined with F#'s sequence expressions.

The following `lines_of_file` function returns a sequence of the lines in the file with the given name:

```
> open System.IO;;
> let lines_of_file filename =
    seq { use stream = File.OpenRead filename
          use reader = new StreamReader(stream)
          while not reader.EndOfStream do
            yield reader.ReadLine() };;
val lines_of_file : string -> seq<string>
```

The sequence expression in this function loops until the end of the file, reading and yielding each line of the file in turn. The `use` bindings are analogous to `let` bindings except that they also dispose of the created objects when enumeration over the sequence is complete. In this case, the `use` bindings cause the file to be closed when enumeration over the sequence is complete.

The .NET platform actually provides similar functions for reading whole files such as `stream.ReadToEnd()` but this sequence expression has an important advantage: the file is only read as the sequence is enumerated over. This is a form of lazy evaluation and is one of the most important benefits of sequences in F#. For example, this `lines_of_file` function can be used to read only the first three lines of a file without having to read the entire file first:

```
> lines_of_file "fib.ml"
  |> Seq.take 3;;
val it : seq<string> =
  ["let rec fib = function";
   "  | 0 | 1 as n -> n";
   "  | n -> fib(n-1) + fib(n-2)"]
```

The `lines_of_file` function can be used to count the number of space-separated words in a file as follows:

```
> lines_of_file @"C:\bible13.txt"
  |> Seq.map (String.split [' '] >> Seq.length)
```

```
|> Seq.fold ( + ) 0;;
val it : int = 823647
```

An important characteristic of this word counter is that it handles a single line at a time and, consequently, has very low memory requirements compared to more obvious solutions that read in the whole file. This kind of programming makes F# interactive sessions a very powerful tool for data dissection and manipulation.

This functionality is of particular importance when handling large amounts of data, such as many database applications. This function will be used to write a simple word frequency counter in chapter 6. The use of sequence expressions to transparently interrogate databases on demand is covered in chapter 10.

5.4 SERIALIZATION

The following save function uses the built-in .NET serialization routines to save *any* value of *any* type to the given file:

```
> open System.Runtime.Serialization.Formatters.Binary;;
> let save filename x =
    use stream =
      new FileStream(filename, FileMode.Create)
    (new BinaryFormatter()).Serialize(stream, x);;
val save : string -> 'a -> unit
```

For example, the following saves a 3-tuple to the "test.dat" file:

```
> save "test.dat" (1, 3.0, ["piece"]);;
val it : unit = ()
```

The following load function reads back a value that was saved by the save function:

```
> let load filename =
    use stream = new FileStream(filename, FileMode.Open)
    (new BinaryFormatter()).Deserialize(stream, x)
    |> unbox;;
val load : string -> 'a
```

For example, the saved 3-tuple can be loaded back with:

```
> (load "test.dat" : int * float * string list);;
val it : int * float * string list = (1, 3.0, ["piece"])
```

Note that we are careful to annotate the type of the data. Although such type annotations can be omitted in compiled code under certain circumstances, it is always a good idea to make this information explicit in the source code as we have done.

The standard .NET serialization routines can store and retrieve arbitrary data in a binary format. Data stored in non-trivial formats is typically read using techniques known as *lexing* and *parsing*. Serialization is simpler but often an order of magnitude slower than using a custom format, even a readable text-based format.

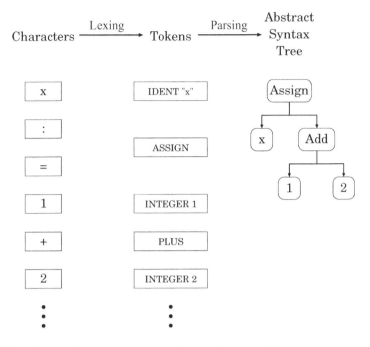

Figure 5.1   Parsing character sequences often entails lexing into a token stream and then parsing to convert patterns of tokens into grammatical constructs represented hierarchically by a tree data structure.

5.5 LEXING AND PARSING

In addition to the primitive input and output functions offered by F#, the language is bundled with very powerful tools, called `fslex` and `fsyacc`, for deciphering the content of files according to a formal grammar. This aspect of the language has been particularly well honed due to the widespread use of this family of languages for writing compilers and interpreters, i.e. programs which understand, and operate on, other programs.

In the context of scientific computing, providing data in a human readable format with a formally defined grammar is highly desirable. This allows a data file to convey useful information both to a human and to a computer audience. In the case of a human, the file can be read using a text editor. In the case of a computer, the file can be *parsed* to produce a data structure which reflects the information in the file (illustrated in figure 5.1). In the latter case, the program could then go on to perform computations on the data.

The ability to use `fslex` and `fsyacc` is, therefore, likely to be of great benefit to scientists. We shall now examine the use of these tools in more detail.

5.5.1   Lexing

The first step in using these tools to interpret a file is called *lexing*. This stage involves reading characters from the input, matching them against patterns and outputting a stream of *tokens*. A token is a value which represents some of the input. For example, a sequence of space-separated digits could be lexed into a stream of integer-valued tokens.

5.5.1.1   Using `String.split`   The task of lexing can be elucidated by the simple example of splitting a file into lines and words, expected to represent the columns and elements of a matrix, respectively.

A function to read a matrix by splitting each line into words, converting each word into a `float` and constructing an F# `Matrix` may be written in terms of the `lines_of_file` function that was defined in section 5.3:

```
> let read_matrix filename =
    lines_of_file filename
    |> Seq.map (String.split [' '])
    |> Seq.map (Seq.map float)
    |> Math.Matrix.of_seq;;
val read_matrix : string -> Math.Matrix
```

Note that reading from the file is deferred until the call to `Matrix.of_seq` because `Seq` is lazy.

The `String.split` function is lazily mapped over this sequence to obtain a sequence of string lists representing the space-separated words on each line. Two sequence maps are used to apply the `float_of_string` function to each element. Finally, the `Matrix.of_seq` function enumerates the elements of the sequences (opening, reading and closing the file) and generates the matrix result.

For example, reading a 3×3 matrix from a file:

```
> read_matrix @"C:\vectors.txt";;
val it : Math.Matrix = matrix [[1.0; 2.0; 3.0];
                               [4.0; 5.0; 6.0];
                               [7.0; 8.0; 9.0];]
```

Splitting strings can be a useful way to lex and parse very simple formats. However, this approach is not very powerful or extensible. In particular, lexing mathematical expressions with a sophisticated syntax would be very tedious. Fortunately, the `fslex` tool allows more complicated lexing to be performed easily.

5.5.1.2   Using `fslex`   The `fslex` tool is a lexer generator that takes F#-like source code in a file with the suffix ".fsl" that describes a lexer and compiles it into efficient F# source code that implements that lexer, which can then be compiled with the F# compiler as an ordinary source file. The resulting lexer translates an incoming stream of characters into an outgoing stream of *tokens*. A token is a variant type. In order to produce tokens, patterns are spotted in the stream of characters by matching them against *regular expressions*, also known as *regexs*. Like pattern matches, the regexs handled by fslex may contain several kinds of structure:

'x' match the specified, single character.

'a'-'z' match any character from a range.

_ match any single character.

"string" match the given string of characters.

Several other constructs are specific to regexps:

['a' 'c' 'e'] match any regexp in the given set.

[^ 'a' 'c' 'e'] match any regexp not in the given set.

regexp * match zero or more repetitions of a string matching regexp.

regexp + match one or more repetitions of a string matching regexp.

regexp ? match regexp or the empty string.

regexp$_1$ # regexp$_2$ match any string which matches regexp$_1$ and does not match regexp$_2$.

regexp$_1$ | regexp$_2$ match any string which either matches regexp$_1$ or matches regexp$_2$.

regexp$_1$ regexp$_2$ concatenate a string matching regexp$_1$ with a string matching regexp$_2$.

eof match the end of file.

In an fslex file (with the suffix ".fsl"), regular expressions can be named using the let construct. A regular expression named digit that matches any decimal digit may be written:

```
let digit = ['0'-'9']
```

String representations of floating-point numbers are somewhat more adventurous. An initial attempt at a regular expression might match a sequence a digits followed by a full-stop followed by another sequence of digits:

```
let floating = digits+ '.'  digits+
```

This will match 12.3 and 1.23 correctly but will fail to match 123. and .123. These can be matched by splitting the regular expression into two variants, one which allows zero or more digits before the full-stop and one which allows zero or more digits after the full-stop:

```
let floating = digit+ '.'  digit* | digit* '.'  digit+
```

Note that a single '.' does not match this floating regexp.

As we have already seen, a conventional notation (e.g. $1,000,000,000,000 \equiv 1e12$) exists for decimal exponents. The exponent portion of the string ("e12") may be represented by the regexp:

```
let exponent = ['e' 'E'] ['+' '-']?  digit+
```

Thus, a regular expression matching positive floating-point numbers represented as strings may be written:

```
let digit = ['0'-'9']
let mantissa = digit+ '.'  digit* | digit* '.'  digit+
let exponent = ['e' 'E'] ['+' '-']?  digit+
let floating = mantissa exponent?
```

On the basis of these example regular expressions for integer and floating-point number representations, we shall now develop a lexer. A file giving the description of a lexer for fslex has the suffix ".fsl". Although lexer definitions depend upon external declarations, we shall examine the description of the lexer first. Specifically, we shall consider an example file "lexer.fsl":

```
{
  open Parser
  open Lexing
}
let digit = ['0'-'9']
let mantissa = digit+ '.' digit* | digit* '.' digit+
let exponent = ['e' 'E'] ['+' '-']?  digit+
let floating = mantissa exponent?
rule token = parse
| [' ' '\t' '\n'] { token lexbuf }
| floating
| digit+              { NUMBER(float(lexeme lexbuf)) }
| '+'                 { ADD }
| '*'                 { MUL }
| eof                 { EOF }
```

This lexer defines regular expressions for digits and floating point numbers (as a mantissa followed by an optional exponent) and uses these regular expressions to convert a stream of characters into a stream of four different kinds of token: NUMBER, ADD, MUL and EOF.

The lexer contains a single rule called token. The first pattern in this rule matches whitespace (spaces, tabs and newlines) and the corresponding action simply calls the token rule recursively to continue lexing, i.e. it skips all whitespace.

The next two patterns match either the floating regular expression or sequences of one or more digits, converting the matched string (given by applying the Lexing.lexeme function to the lexbuf variable) into a token containing the number as a value of the type float. The order of the regular expressions is important in this case. If the stream were searched for digits first then the front of a floating point number would be matched up to the decimal point or e, leaving either an incorrect float to be parsed next (the fractional part) or an erroneous result. For example, the regular expression digit+ would match 123 in 123.456, leaving .456 to be matched next.

The next two patterns match the + and * symbols in the input, generating tokens called ADD and MUL, respectively.

The final pattern generates an EOF token when the input character stream ends (denoted by the built-in regexp eof).

This lexer can be compiled into an F# program using the fslex compiler which, at the time of writing, must be invoked manually from a DOS prompt. For example:

```
C:\...> fslex lexer.fsl
compiling to dfas (can take a while...)
148 NFA nodes
17 states
writing output
```

The resulting F# source code which implements a lexer of this description is placed in the "lexer.fs" file. Before this file can be compiled, we must create the Parser module which it depends upon.

In many cases, a parser using a lexer would itself be generated from a parser description, using the fsyacc compiler. We shall describe this approach in the next section but, before this, we shall demonstrate how the functionality of a generated lexer may be exploited without using fsyacc.

Before compiling the F# program "lexer.fs", which implements our lexer, we must create the Parser module which it depends upon:

```
> module Parser =
    type token =
      | NUMBER of float
      | ADD
      | MUL
      | EOF;;
```

Note that the NUMBER, ADD, MUL and EOF tokens used by the lexer are actually nothing more than type constructors, in this case for for the Parser.token type. Having defined the Parser module and, in particular, the Parser.token variant type, we can include the Lexer module in the interactive session using the #load directive:

```
> #load "lexer.ml";;
...
```

A function to lex a string into a sequence of tokens may be written:

```
> let rec lex lexbuf =
    match Lexer.token lexbuf with
```

```
    | Parser.EOF as token -> [token]
    | token -> token :: lex lexbuf;;
val lex :
  Tools.FsLex.LexBuffer<Lexing.position, byte> ->
    Parser.token list
```

For example, the expression $1 + 2 \times 3$ is lexed into six tokens:

```
> lex (Lexing.from_string "1 + 2 * 3");;
val it : seq<token>
  = [NUMBER 1.0; ADD; NUMBER 2.0; MUL; NUMBER 3.0; EOF]
```

The capabilities of a lexer can clearly be useful in a stand-alone configuration. In particular, programs using lexers, such as that we have just described, will validate their input to some degree. In contrast, many current scientific applications silently produce erroneous results. However, the capabilities of a lexer can be greatly supplemented by an associated parser, as we shall now demonstrate.

5.5.2 Parsing

The parsing stage of interpreting input converts the sequence of tokens from a lexer into a hierarchical representation (illustrated in figure 5.1) - the *abstract syntax tree* (**AST**). This is performed by accumulating tokens from the lexer either until a valid piece of grammar is recognised and can be acted upon, or until the sequence of tokens is clearly invalid, in which case a `Parsing.Parse_error` exception is raised.

The `fsyacc` compiler can be used to create F# programs which implement a specified parser. The specification for a parser is given by a description in a file with the suffix ".fsy". Formally, these parsers implement LALR(1) grammars described by rules provided in *Backus-Naur form* (**BNF**).

For example, consider a parser, based upon the lexer described in the previous section, which interprets textual data files. These files are expected to contain simple mathematical expressions composed of numbers, additions and multiplications.

A program to parse these files can be generated by `fsyacc` from a grammar which we shall now describe. Given a file "*name*.fsy" describing a grammar, by default `fsyacc` produces a file "*name*.fs" containing an F# program implementing that grammar. Therefore, in order to generate a `Parser` module for the lexer, we shall place our grammar description in a "parser.fsy" file.

Grammar description files begin by listing the definitions of the tokens which the lexer may generate. In this case, the possible tokens are NUMBER, ADD and MUL and EOF.

Tokens which carry no associated values are defined with:

`%token` *token*$_1$ *token*$_2$ ...

Tokens which carry an associated value of type *type* are defined by:

`%token` *<type> token*$_1$ *token*$_2$ ...

Thus, the tokens for our lexer can be defined by:

```
%token ADD MUL EOF
%token <float> NUMBER
```

The token definitions are followed by declarations of the associativities and relative precedences (listed in order of increasing precedence) of tokens. In this case, both ADD and MUL are left associative and MUL has higher precedence:

```
%left ADD
%left MUL
```

Then a declaration of the entry point into the parser. In this case, we shall use a parsing rule called expr to parse expressions:

```
%start expr
```

This is followed by a declaration of the type returned by the entry point. In this case, we shall use a type expr:

```
%type <expr> expr
```

Before the rules and corresponding actions describing the grammar and parser are given, the token and entry-point definitions are followed by a separator:

```
%%
```

Like ordinary pattern matching, the guts of a parser is represented by a sequence of *groupings* of *rules* and their corresponding *actions*:

group :
| *rule*$_1$ { *action*$_1$ }
| *rule*$_2$ { *action*$_2$ }
. . .
| *rule*$_n$ { *action*$_n$ }
;

A grouping represents several possible grammatical constructs, all of which are used to produce F# values of the same type, i.e. the expressions *action*$_1$ to *action*$_n$ must have the same type. Rules are simply a list of expected tokens and groupings. In particular, rules may be recursive, i.e. they may refer to themselves, which is useful when building up arbitrarily long lists.

Tokens and rule groups carry values. The value of the nth-matched subexpression of a rule is referred to by the variable $\$n$ in the corresponding action.

This example parser contains a single, recursive group of rules that parses simple symbolic expressions:

```
expr:
| NUMBER { Num $1 }
| OPEN expr CLOSE { $2 }
| expr ADD expr { $1 + $3 }
| expr MUL expr { $1 * $3 }
;
```

The first rule matches a NUMBER token and generates a Num node in the AST with the float value carried by the token (denoted by $1). The second rule returns the AST of a bracketed expression. The third and fourth rules handle the infix addition and multiplication operators, generating Add and Mul nodes in the AST using the static member functions of the expr type.

An F# program "parser.fs" and interface "parser.fsi" that implement this parser, described in this "parser.fsy" file, may be compiled using the fsyacc program:

```
$ fsyacc parser.fsy
building tables
computing first function
building kernels
computing lookahead relations
building lookahead table
building action table
building goto table
10 states
2 nonterminals
9 terminals
5 productions
#rows in action table: 10
#unique rows in action table: 7
maximum #different actions per state: 4
average #different actions per state: 2
```

The lexer and parser can be tested from the F# interactive mode. The parser must be loaded before the lexer and the expr type must be loaded before the parser. Therefore, we begin with the definition of the expr type:

```
> type expr =
    | Num of float
    | Add of expr * expr
    | Mul of expr * expr
  with
    static member ( + ) (f, g) = Add(f, g)
    static member ( * ) (f, g) = Mul(f, g);;
```

The Parser module that uses this expr type can be loaded from the generated file using the #load directive:

```
> #load "parser.fs";;
...
```

The Lexer module that uses the Parser module can be loaded from the generated file using the #load directive:

```
> #load "lexer.fs";;
...
```

Finally the lexer and parser can be used to parse a simple symbolic expression to produce an abstract syntax tree by applying the `Lexer.token` function and the lex buffer to the higher-order `Parser.expr` function:

```
> Lexing.from_string "1 + 2 * 3"
  |> Parser.expr Lexer.token;;
val it : expr = Add(Num 1.0, Mul(Num 2.0, Num 3.0))
```

Note that the lexer handled the whitespace and the parser handled the operator precedences.

Thus our lexer and parser have worked together to interpret the integer and floating-point numbers contained in the given text as well as the structure of the text in order to convert the information contained in the text into a data structure that can then be manipulated by further computation.

Moreover, the lexer and parser help to validate input. For example, input that erroneously contains letters is caught by the lexer. In this case, the lexer will raise `FailureException`. A file which erroneously contains a floating-point value where an integer was expected will be lexed into tokens without fault but the parser will spot the grammatical error and raise the `Parsing.Parser_error` exception.

In an application, the call to the `main` function in the `Parser` module can be wrapped in a `try..with` to catch the `Parsing.Parser_error` exception and handle it.

As we have seen, the `fslex` and `fsyacc` compilers can be indispensable in both designing and implementing programs that use a non-trivial file format. These tools are likely to be of great benefit to scientists wishing to create unambiguous, human-readable formats.

CHAPTER 6

SIMPLE EXAMPLES

This chapter presents a wide variety of code snippets implementing useful functions in different ways. These examples should serve to ossify the readers understanding of the F# language in order to prepare them for the complete example programs in chapter 12. Whilst describing these functions we endeavour to explain the relationships between the styles chosen and the information disseminated in the previous chapters of this book.

6.1 FUNCTIONAL

Many useful mathematical and scientific constructs can be encoded very elegantly by exploiting the features of functional programming. In the introduction, we saw a remarkably simple and yet powerful numerical derivative function. In this section, we shall examine more functions that will be useful to scientists using F#.

F# for Scientists. By Jon Harrop
Copyright © 2008 John Wiley & Sons, Inc.

6.1.1 Nest

The nest function is found in many technical computing environments. This combinator nests applications of its function argument f to a given value x a given number of times n. The nest function may be written:

```
> let rec nest n f x =
    match n with
    | 0 -> x
    | n -> nest (n - 1) f (f x);;
val nest : int -> ('a -> 'a) -> 'a -> 'a
```

Note that we have phrased the recursive call to nest such that it is tail recursive. An equivalent but non-tail-recursive implementation would have been f(nest (n - 1) f x).

For example, in the context of the numerical derivative combinator d and the example function $f = x^3 - x - 1$ from section 1.6.4, the third derivative of f may be written:

```
> let f''' = nest 3 d f;;
val f''' : (float -> float)
```

The nest combinator has many uses.

6.1.2 Fixed point

Iterative algorithms sometimes terminate when an iteration leaves the result unchanged. This is known as *iterating to fixed point*. A fixed_point combinator can implement this functionality using recursion:

```
> let rec fixed_point f x =
    match f x with
    | f_x when x = f_x -> x
    | x -> fixed_point f x;;
val fixed_point : ('a -> 'b) -> 'a -> 'b
```

For example, the golden ratio may be expressed as the fixed point of the iteration $x_{n+1} = \sqrt{1 + x_n}$:

```
> fixed_point (fun x -> sqrt (1.0 + x)) 1.0;;
val it : float = 1.618033989
```

Note that this iterative algorithm does not terminate in theory. The function only terminated because the next result was rounded to the same float value, i.e. the result was unchanged to within the error of a floating point value.

6.1.3 Within

In scientific computing, many algorithms act only upon values that lie within a specific range. For example, to sum the elements of a list that lie within the range $[3 \ldots 6]$ we might filter out those elements first:

```
> [1 .. 10]
  |> List.filter (fun x y -> 3 <= y && y <= 6)
  |> List.fold_left (+) 0;;
val it : int = 18
```

or filter as we fold by specifying a more complicated function argument:

```
> [1 .. 10]
  |> List.fold_left
        (fun x y ->
          if 3 <= y && y <= 6 then x + y else x) 0;;
val it : int = 18
```

The former approach is less efficient because it allocates and then deallocates a temporary list. The latter approach is more efficient but less comprehensible. Consequently, it is useful to have a combinator that "propagates" a fold only when the argument lies within a specified range:

```
> let within x0 x1 f accu x = combinator
    if x0 <= x && x <= x1 then f accu x else accu;;
val within :
  'a -> 'a -> ('b -> 'a -> 'b) -> 'b -> 'a -> 'b
```

For example, the `within` combinator can be used to filter out elements lying within a range without creating an intermediate data structure or resorting to sequences (which are also slower):

```
> List.fold_left (within 3 6 ( + )) 0 [1 .. 10];;
val it : int = 18
```

Note that we have been careful to order the arguments to the `within` function such that currying can be exploited to simplify uses of this function in a left fold.

Higher-order functions can clearly be used to aggressively factor code without sacrificing clarity or performance. Indeed, clarity has arguably been improved in this case and, as we shall see in chapter 8, this is an excellent way to write efficient code.

6.1.4 Memoize

Caching the results of computations to avoid recomputation can be a very effective optimization. This applies not only to complicated numerical computations but also to functions that are slowed by external limitations, such as database connectivity, web access and so on.

The Fibonacci function is the pedagogical example of a function that can be accelerated using caching:

```
> let rec fib = function
    | 0 | 1 as n -> n
    | n -> fib(n - 1) + fib(n - 2);;
val fib : int -> int
```

This function is slow to compute even small Fibonacci numbers, taking 0.222s to compute the 35th Fibonacci number:

```
> time fib 35;;
Took 222ms
val it : int = 9227465
```

The fib function is heavily recursive, making two recursive calls almost every time it is invoked (each of which make two more recursive calls and so on). Consequently, the asymptotic complexity of this function is $O(2^n)$. As the function is slow to execute, the results can be productively cached for reuse.

6.1.4.1   Simple memoization   Fortunately, even the generic concept of caching the results of function calls can be factored out and expressed as a higher-order function. This technique is known as *memoization*. A higher-order function to memoize a given function may be written:

```
> let memoize f =
    let m = Hashtbl.create 1
    fun x ->
      try m.[x] with Not_found ->
      let f_x = f x
      m.[x] <- f_x
      f_x;;
val memoize : ('a -> 'b) -> 'a -> 'b
```

This memoize combinator can be applied to the fib function to produce a new function that has the same type as the fib function but transparently caches previously computed results:

```
> let mfib = memoize fib;;
val mfib: (int -> int)
```

This function is slow when invoked for the first time on any given input but is fast the next time it is invoked with the same input:

```
> time mfib 35;;
Took 240ms
val it : int = 9227465
> time mfib 35;;
Took 0ms
val it : int = 9227465
```

Memoization can clearly be useful in a variety of circumstances. However, this implementation of the memoize combinator can be generalized even further.

6.1.4.2   Recursive memoization   The previous example cached the return values of the original fib function to improve performance when a calculation was repeated. However, when computing the 35th Fibonacci number, the function still recomputes the 33rd and 34th Fibonacci numbers. Caching would be more effective if it could memoize the results of the recursive calls made inside the fib function as well as the calls made from the outside.

This can be done by first unravelling the fib function such that it is no longer recursive, a technique known as "untying the recursive knot". To do this, the function is written as a higher-order function and is passed the function that it will recurse into. This is most simply written by removing the rec from the definition and adding an initial argument with the same name as the function itself:

```
> let fib fib = function
    | 0 | 1 as n -> n
    | n -> fib(n - 1) + fib(n - 2);;
val fib : (int -> int) -> int -> int
```

By passing this fib function a memoized version of itself as its first argument, the recursive calls inside the fib function can also be memoized. Doing this required a slightly modified version of the memoize combinator as well:

```
> let memoize_rec f =
    let m = Hashtbl.create 1
    let rec f' x =
      try m.[x] with Not_found ->
      let f_x = f f' x
      m.[x] <- f_x
      f_x in
    f';;
val memoize : (('a -> 'b) -> 'a -> 'b) -> 'a -> 'b
```

The nested f' function is the memoized version of the f function. Note that the f' function is careful to pass itself as the first argument to f, so that calls in f that were recursive become calls to the memoized version f' instead.

Applying memoize_rec to the untied fib function creates a recursively memoized version of the fib function:

```
> let mfib = memoize_rec fib;;
val mfib : (int -> int)
```

This function is not only quick to recompute Fibonacci numbers but is now asymptotically quicker to compute all Fibonacci numbers! Specifically, computing the nth number only requires the previous n numbers $[0 \ldots n - 1]$ to be computed, i.e. the asymptotic complexity is now only $O(n)$. Consequently, this function can be used to compute larger Fibonacci numbers very quickly.

For example, computing the 45th Fibonacci number takes only 2ms:

```
> time mfib 45;;
Took 2ms
val it : int = 1134903170
```

The ability to represent memoization as a combinator makes it much easier to use in functional languages. However, the F# programming language actually provides sophisticated support for recursive definitions and, in fact, can represent this recursive memoization using only the original `memoize` function:

```
> let rec fib =
    memoize
      (function
         | 0 | 1 as n -> n
         | n -> fib(n - 1) + fib(n - 2));;
val fib : (int -> int) -> int -> int
```

The compiler will emit a warning explaining that the soundness of this function definition cannot be checked at compile time and run-time checks will be inserted automatically but we can prove to ourselves that the definition is correct. This approach elegantly combines the brevity and clarity of the original implementation with the efficiency of a completely memoized function, including recursive calls:

```
> time mfib 45;;
Took 2ms
val it : int = 1134903170
```

However, the growth of the hash table used to memoize previously-computed results in this implementation can grow unbounded. This can sometimes be an effective memory leak when previous results are no longer needed and should be "forgotten". Consequently, it may be useful to replace the memoize combinator with an implementation that uses a more sophisticated data structure to remember only a certain number of the most commonly used answers, or even to measure the memory required by memoization and limit that.

6.1.4.3 *Dynamic programming*

Divide and conquer describes the strategy of algorithms that break large problems into separate smaller problems. This is a vitally important topic in algorithm design and underpins many high-performance algorithms that make practical problems tractable. A closely related subject that has received growing attention in recent years is called "dynamic programming". This refers to strategies similar to divide and conquer that manage to break large problems into *overlapping* smaller problems. Optimal asymptotic efficiency is then most easily obtained by memoizing the overlaps and reusing them rather than recomputing them.

The Fibonacci function described in the previous section is perhaps the simplest example of dynamic programming. At each step, the classic description of the Fibonacci function breaks the large computation of `fib n` into two smaller computations of `fib(n - 1)` and `fib(n - 2)` but the smaller computations overlap because `fib(n - 1)` can be expressed in terms of `fib(n - 2)` and `fib(n - 3)`. Applying the `memoize` combinator, as we did, allows the overlapping solutions to be reused and greatly improves performance as a consequence.

Dynamic programming will be revisited in section 12.3.

6.1.5 Binary search

Many problems can be solved using recursive bisection and the generic algorithm is known as *binary search*.

The following function splits a range that is assumed to bracket a change in the comparison function cmp and returns the subrange that brackets the change:

```
> let binary_search split cmp (x, c_x, y, c_y) =
    let m = split x y
    let c_m = cmp m
    match c_x = c_m, c_m = c_y with
    | true, false -> m, c_m, y, c_y
    | false, true -> x, c_x, m, c_m
    | _ -> raise Not_found;;
val binary_search :
    ('a -> 'a -> 'a) -> ('a -> 'b) -> 'a * 'b * 'a * 'b ->
    'a * 'b * 'a * 'b
```

Note that this function is careful to avoid recomputation of cmp with the same argument, in case this is a slow function. This is achieved by bundling the values of cmp x and cmp y in the range and reusing them in the subrange that is returned.

The choice of currying the function arguments split and cmp but coalescing the range into a 4-tuple might seem odd but this is, in fact, a careful design decision. Thanks to this design, the binary_search function can be partially applied with its function arguments to yield a function that maps ranges onto ranges. Consequently, such a partial application can then be used with combinators such as nest and fixed_point to control the recursive subdivision of ranges.

This function will be used in the context of root finding, in section 6.2.5.

6.2 NUMERICAL

Many useful functions simply perform computations on the built-in number types. In this section, we shall examine progressively more sophisticated numerical computations.

6.2.1 Heaviside step

The heaviside step function:

$$H(x) = \begin{cases} 0 & x < 0 \\ 1 & x \geq 0 \end{cases}$$

is a simple numerical function which may be implemented trivially in F#:

```
> let heaviside x =
    if x < 0.0 then 0.0 else 1.0;;
val heaviside : float -> float
```

This function is particularly useful when composed with other functions.

6.2.2 Kronecker δ-function

The Kronecker δ-function:

$$\delta_{ij} = \begin{cases} 1 & i = j \\ 0 & i \neq j \end{cases}$$

may be written as:

```
> let kronecker i j =
    if i = j then 1 else 0;;
val kronecker : 'a -> 'a -> int
```

However, this implementation is polymorphic, which may be undesirable because static type checking will not enforce the application of this function to integer types only. Consequently, we may wish to restrict the type of this function by adding an explicit type annotation:

```
> let kronecker (i : int) j =
    if i = j then 1 else 0;;
val kronecker : int -> int -> int
```

Erroneous applications of this function to inappropriate types will now be caught at compile-time by the F# compiler.

6.2.3 Gaussian

Computations involving trigonometric functions may be performed using the `sin`, `cos`, `tan`, `asin` (arcsin), `acos` (arccos), `atan` (arctan), `atan2` (as for `atan` but accepting signed numerator and denominator to determine which quadrant the result lies in), `cosh`, `sinh` and `tanh` functions.

The conventional mathematical functions $\sqrt{x}$ (`sqrt x`) and e^x (`exp x`) are required to compute the Gaussian:

$$f(x) = \frac{1}{\sqrt{2\pi}\sigma} e^{-\frac{1}{2\sigma^2}(x-\mu)^2}$$

Thus, a function to calculate the Gaussian may be written:

```
> let gaussian mu sigma (x : float) =
    let sqr x = x * x
    exp(- sqr(x - mu) / (2.0 * sqr sigma)) /
      (sqrt(2.0 * System.Math.PI) * sigma)
val gaussian : float -> float -> float -> float
```

As this implementation of the Gaussian function is curried, a function representing a probability distribution with a given μ and σ may be obtained by partially applying the first two arguments:

```
> seq { for x in -1.0 .. 0.5 .. 1.0 ->
          gaussian 1.0 0.5 x };;
```

Figure 6.1 The first seven rows of Pascal's triangle.

```
val it : seq<float> =
  seq [0.0002676604; 0.008863696; 0.107981933;
       0.483941449; ...]
```

Many of the special functions can be productively written in curried form to allow their arguments to be partially applied.

6.2.4 Binomial coefficients

The binomial coefficient $\binom{n}{r}$ is typically defined in mathematics as:

$$\binom{n}{r} = \frac{n!}{r!(n-r)!}$$

Naturally, this may be written directly in terms of the factorial function:

```
> let binomial n r =
    factorial n / (factorial r * factorial (n - r));;
val binomial bigint -> bigint -> bigint
```

For example, the binomial coefficients in the expansion of $(a + b)^6$ are given by:

```
> [ for r in 0I .. 6I ->
      binomial 6I r ];;
val it : bigint list = [1I; 6I; 15I; 20I; 15I; 6I; 1I]
```

However, computing relatively small binomial coefficients requires the computation of large intermediate values due to the use of factorials. For example, the computation of $\binom{1000}{2} = 499500$ requires the 2,568-digit value of 1000! to be computed and then discarded, which takes around 12ms:

```
> time (loop 1000 (binomial 1000I)) 2I;;
Took 12123ms
val it : bigint = 499500I
```

This implementation of the binomial function is clearly a suitable candidate for algorithmic optimization.

The performance of this binomial function is most easily improved upon by computing Pascal's triangle, where each number in the triangle is the sum of its two "parents" from the row before (illustrated in figure 6.1). This may be represented as the recurrence relation:

$$\binom{n}{r} = \begin{cases} 1 & r = 0 \\ 1 & r = n \\ \binom{n-1}{r} + \binom{n-1}{r-1} & \text{otherwise} \end{cases}$$

When using finite precision arithmetic (e.g. the int type), computing binomial coefficients using Pascal's triangle is more robust than computing via factorials because the numbers involved now increase monotonically, only overflowing if the result overflows. Our example can then be computed using only machine precision integers, even on 32-bit machines.

The recurrence relation may be expressed as a recursive function:

```
> let rec binomial n r =
    if r = 0 || r = n then 1 else
    binomial (n - 1) r + binomial (n - 1) (r - 1);;
val binomial : int -> int -> int
```

This implementation of the binomial function is slightly faster than the factorial-based implementation, taking around 9ms to compute $\binom{1000}{2}$ using finite-precision integer arithmetic:

```
> time (loop 1000 (binomial 1000)) 2;;
Took 8904ms
val it : int = 499500
```

Moreover, this is a divide and conquer algorithm amenable to memoization. Consequently, this binomial function is most easily optimized using the memoize combinator from section 6.1.4.1, replacing the two curried arguments n and r with a single 2-tuple argument:

```
> let rec binomial =
    memoize
      (fun (n, r) ->
         if r = 0 || r = n then 1 else
         binomial(n - 1, r) + binomial(n - 1, r - 1));;
val binomial : (int * int -> int) -> int * int -> int
```

The resulting function is $65\times$ faster than the first implementation at the same computation:

```
> time (loop 1000 (binomial 1000)) 2;;
Took 188ms
val it : int = 499500
```

Let us examine some more sophisticated numerical functions.

6.2.5   Root finding

The fixed_point and binary_search combinators from sections 6.1.2 and 6.1.5, respectively, can be used to implement a simple root finder by recursively subdividing a range that brackets the zero-crossing of a function:

```
> let find_root f x y =
    let split x y = (x + y) / 2.0
    let cmp x = compare (f x) 0.0
    (x, cmp x, y, cmp y)
    |> fixed_point (binary_search mid cmp);;
val find_root :
  (float -> float) -> float -> float ->
    float * int * float * int
```

For example, this function can be used to find the root of the function $f(x) = x^3 - x - 1$ that lies in the range $x \in \{1 \ldots 2\}$:

```
> find_root (fun x -> x**3.0 - x - 1.0) 1.0 2.0;;
val it : float * int * float * int =
  (1.324717957, -1, 1.324717957, 1)
```

This result contains two very similar values of x that bracket the root.

When writing programs to perform scientific computations, programmers should constantly consider the possibility of factoring out higher-order functions. Although functional programming languages are becoming increasingly popular, the vast majority of the existing literature covers fundamental algorithms written in a relatively unfactored and typically very imperative style.

6.2.6   Grad

The ∇ operator in mathematics computes the vector derivative of a function of many variables:

$$\nabla f(x, y, z, \ldots) = \left(\frac{\partial f}{\partial x}, \frac{\partial f}{\partial y}, \frac{\partial f}{\partial z}, \ldots \right)$$

In F#, a function of many variables may be represented by the type:

```
val f : vector -> float
```

The vector value of ∇f may be computed as the numerical derivative of f with respect to each variable using the same procedure as the one-dimensional numerical derivative function d from section 1.6.4.

A numerical approximation to the partial derivative of f with respect to its ith variable may be calculated using the following function:

```
> let partial_d f_xs f xs i xi =
    xs.[i] <- xi + delta
    try (f xs - f_xs) / delta finally
    xs.[i] <- xi;;
```

```
val d :
  float -> (vector -> float) -> vector -> int ->
    float -> float
```

where f_xs is the value of f xs and xi is the original value of x_i. This definition of the partial_d function allows f_xs, f and xs to be partially applied such that the resulting closure can be applied to the conventional mapi function to obtain the grad.

Note the use of the try ... finally construct to ensure that the state change to the array xs is undone even if the application f xs causes an exception to be raised.

The vector value of ∇f can then be computed using a simple combinator:

```
> let grad f xs =
    Vector.mapi (partial_d (f xs) f xs) xs;;
val grad : (vector -> float) -> vector -> vector
```

The efficiency of this grad function is likely to be dominated by the cost of applying f. This implementation has factored out the computation of f xs for the original xs, so this grad function only applies the f function $T(n + 1)$ times.

This grad function can be used in many important numerical algorithms, such as the minimization of multidimensional functions.

6.2.7 Function minimization

The task of altering the variables of a function in order to minimize or maximize the value of the function is common in scientific computing. By convention, this class of problems are generally referred to as minimization problems. There are many algorithms for function minimization and they are broadly classified into local and global minimization. Local function minimization refers to the task of tweaking the variables from an initial set in order to find a maximum that is close to the initial values by monotonically decreasing the value of the function. In contrast, global function minimization refers to grossly altering the variables in an attempt to find the highest possible value of the function, typically allowing the value of the function to increase in the interim period.

One of the simplest local function maximization algorithms is called *gradient descent*. This algorithm repeatedly steps in the direction of the downward gradient $-\nabla f$ to monotonically decrease the value of the function f. In order to achieve good performance, the step size λ is increased when the function is successfully decreased and the step is accepted and λ is drastically decreased if the function increases:

$$\lambda_{n+1} = \begin{cases} \alpha\lambda_n & f(\mathbf{x}_n - \lambda\nabla f) \geq f(\mathbf{x}_n) \\ \beta\lambda_n & f(\mathbf{x}_n - \lambda\nabla f) < f(\mathbf{x}_n) \end{cases}$$

for some $\alpha \ll 1$ and $\beta > 1$.

```
> let descend alpha beta f f'
    (lambda, xs: vector, f_xs) =
```

```
    let xs_2 = xs - lambda $* f' xs
    let f_xs_2 = f xs_2
    if f_xs_2 >= f_xs then
      alpha * lambda, xs, f_xs
    else
      beta * lambda, xs_2, f_xs_2;;
val descend :
  float -> float -> (vector -> 'a) ->
    (vector -> vector) -> float * vector * 'a ->
      float * vector * 'a
```

Note that this descend function is designed to work with the fixed_point combinator from section 6.1.2 after partial application of alpha, beta, f, and f'.

In the interests of numerical robustness, the algorithm is careful to decrease λ if the function stays the same to within numerical error. When numerical accuracy is exhausted, $f(\mathbf{x})$ stops changing and λ is monotonically decreased until it underflows to 0 at which point the accumulator $(\lambda, \mathbf{x}, f(\mathbf{x}))$ stops changing and the fixed_point combinator terminates.

The gradient descent algorithm applies the fixed_point combinator to this auxiliary function descend with suitable arguments and extracting the vector of variables x from the result:

```
> let gradient_descent f f' xs =
    let _, xs, _ =
      fixed_point (descend 0.5 1.1 f f') (1.0, xs, f xs)
    xs;;
val gradient_descent :
  (vector -> 'a) -> (vector -> vector) -> vector ->
    vector
```

For example, consider the function $f(x, y) = x^4 + y^2 - x^3 y - 3x$:

```
> let f v =
    let x, y = v.[0], v.[1]
    x**4.0 + y**2.0 - x**3.0 * y - 3.0 * x;;
val f : vector -> float
```

A local minimum of $f(x, y)$ starting from $(x, y) = (0, 0)$ is most easily found using the numerical grad function from section 6.2.6:

```
> gradient_descent f (grad f) (vector [0.0; 0.0]);;
val it : vector = [1.1274523; 0.716579731]
```

Many important algorithms in scientific computing can be composed from simpler functions using higher-order functions, currying and combinators.

6.2.8 Gamma function

Numerical computing often requires the definition of various special functions. Several useful functions such as $\sin(x)$ and $\arctan(x)$ are built-in but many more functions must be defined in terms of these. The gamma function:

$$\Gamma(z) = \int_0^\infty t^{z-1}e^{-t}dt$$

This function arises particularly in the context of factorial functions because it is a generalization of factorial:

$$n! = \Gamma(n+1)$$

Also, the gamma function satisfies the following recurrence relation:

$$\Gamma(z+1) = z\Gamma(z)$$

A good numerical approximation to $\Gamma(z)$ for $\Re[z] > 0$ may be derived from Lanczos' approximation [18]:

$$\Gamma(z) = \frac{\sum_{n=0}^{N} q_n z^n}{\prod_{n=0}^{N}(z+n)}(z+5.5)^{z+0.5}e^{-(z+5.5)}$$

Using the first six terms gives the following implementation of $\ln(\Gamma(x))$:

```
> let q =
    [| 75122.6331530; 80916.6278952; 36308.2951477;
       8687.24529705; 1168.92649479; 83.8676043424;
       2.50662827511 |]
    |> Array.map (fun x -> complex x 0.0);;
val q : Math.complex array
> let gamma z =
    let f x = complex x 0.0
    let z2 = z * z
    let z4 = z2 * z2
    let pow z1 z2 = exp(z2 * log z1)
    (q.[0] + q.[1] * z + q.[2] * z2 + q.[3] * z * z2 +
     q.[4] * z4 + q.[5] * z * z4 + q.[6] * z2 * z4) /
     (z * (f 1.0 + z) * (f 2.0 + z) * (f 3.0 + z) *
      (f 4.0 + z) * (f 5.0 + z) * (f 6.0 + z)) *
     pow (z + f 5.5) (z + f 0.5) * exp(-(z + f 5.5)));;
val gamma : Math.complex -> Math.complex
```

We can test this function on some well-known identities involving the gamma function. For example, $\Gamma(6) = 5! = 120$:

```
> gamma (complex 6.0 0.0);;
val it : Math.complex = 120.0r + 0.0i
```

The following identity also holds:

$$\Gamma\left(\frac{1}{3}\right) \times \Gamma\left(\frac{2}{3}\right) = \frac{2\pi}{\sqrt{3}}$$

```
> gamma(complex (1.0 / 3.0) 0.0) *
    gamma(complex (2.0 / 3.0) 0.0);;
val it : Math.Complex = 3.627598728r+0.0i
> 2.0 * System.Math.PI / sqrt 3.0;;
val it : float = 3.627598728
```

This and other similar functions have a wide variety of uses in scientific computing.

6.2.9   Discrete wavelet transform

Wavelet transforms have many applications in science and engineering. These applications rely largely upon the on the unique time-frequency properties of this class of transforms. Wavelet transforms are most broadly classified into continuous and discrete wavelet transforms. Continuous wavelet transforms are widely used in the sciences and engineering for signal analysis, most notably time-frequency analysis. Discrete wavelet transforms are widely used in computer science and information theory for signal coding, particularly as a way of preconditioning signals to make them more amenable to compression.

All wavelet transforms consider their input (taken to be a function of time) in terms of oscillating functions (wavelets) that are localised in terms of both time and frequency. Specifically, wavelet transforms compute the inner product of the input with child wavelets which are translated dilates of a *mother wavelet*. As the mother wavelet is both temporally and spectrally localised, the child wavelets (as dilated translates) are distributed over the time-frequency plane. Thus, the wavelet transform of a signal conveys both temporal and spectral content simultaneously. This property underpins the utility of wavelets.

Discrete wavelet transforms of a length n input restrict the translation and dilation parameters to n discrete values. Typically, the mother wavelet is defined such that the resulting child wavelets form an orthogonal basis. In 1989, Ingrid Daubechies introduced a particularly elegant construction which allows progressively finer scale child wavelets to be derived via a recurrence relation [4]. This formulation restricts the wavelet to a finite width, a property known as *compact support*. In particular, the *pyramidal* algorithm [20, 21] implementing Daubechies' transform (used by the above functions) requires only $O(n)$ time complexity, even faster than the FFT. The Haar wavelet transform is the simplest such wavelet transform.

In this section, we shall examine a simple form of wavelet transform known as the Haar wavelet transform. Remarkably, the definition of this transform is more comprehensible when given as a program, rather than as a mathematical formulation or English description.

The Haar wavelet transform of a length $n = 2^p\, p \geq 0 \in \mathbb{Z}$ float list is given by the following function:

```
> let rec haar_aux (xs : float list) ss ds =
    match xs, ss, ds with
    | [ss], [], ds -> ss::ds
    | [], ss, ds -> haar_aux ss [] ds
    | x1::x2::xs, ss, ds ->
        haar_aux xs (x1 + x2 :: ss) (x1 - x2 :: ds)
    | _ -> invalid_arg "haar";;
val haar_aux :
  float list -> float list -> float list -> float list
> let haar xs =
    haar_aux xs [] [];;
val haar : float list -> float list
```

For example, the Haar wavelet transform converts the following sequence into a more redundant sequence that is more amenable to other data compression techniques:

```
> haar [1.0; 2.0; 3.0; 4.0; -4.0; -3.0; -2.0; -1.0];;
val it : float list =
  [0.0; 20.0; 4.0; 4.0; -1.0; -1.0; -1.0; -1.0]
```

The ihaar_aux function implements the transform by tail recursively taking pairs of elements off the input list and prepending the sum and difference of each pair onto two internal lists called ss and ds, respectively. When the input is exhausted, the process is repeated using the list of sums of pairs as the new input. Finally, when the input contains only a single element, the result is obtained by prepending this element (the total sum) onto the list of differences.

The inverse transform may be written:

```
> let rec ihaar_aux xs ss ds =
    match xs, ss, ds with
    | xs, [], [] -> xs
    | ss, [], ds -> ihaar_aux [] ss ds
    | xs, x1::ss, x2::ds ->
        ihaar_aux (0.5 * (h1+h2) :: 0.5 * (x1-x2) :: xs)
          ss ds
    | _ -> invalid_arg "ihaar"
val haar_aux :
  float list -> float list -> float list -> float list
> let ihaar = function
    | [] -> []
    | ss::ds -> ihaar_aux [] [ss] ds;;
val ihaar : float list -> float list
```

The previous example transform can be reversed to recover the original data from its representation in the wavelet basis:

```
> ihaar [0.0; 20.0; 4.0; 4.0; -1.0; -1.0; -1.0; -1.0];;
val it : float list =
```

```
[1.0; 2.0; 3.0; 4.0; -4.0; -3.0; -2.0; -1.0]
```

Wavelet transforms are often perceived as a complicated form of analysis but, as these example functions have shown, discrete wavelet transforms can be implemented quickly and easily in F#.

6.3 STRING RELATED

In recent times, the dominance of numerical methods in scientific computing has been significantly displaced by other forms of analysis. Particularly in the context of DNA and protein sequences, scientists are analysing a wider range of data structures than before. Strings are commonly used to represent such sequences and, consequently, are commonly using in bioinformatics as well as more directly-related sciences such as computational linguistics.

6.3.1 Transcribing DNA

Transcription creates a single-strand RNA molecule from the double-strand DNA. The process replaces the nucleic acid thymine with uracil.

Consider the representation of a DNA sequence as a string containing the characters A, C, G and T:

```
> let dna = "ACGTTGCAACGTTGCAACGTTGCA";;
val dna : string
```

The `String.map` function is the simplest way to replace single characters in a string. For example, the transcription of dna is given by:

```
> String.map (function 'T' -> 'U' | c -> c) dna;;
val it : string = "ACGUUGCAACGUUGCAACGUUGCA"
```

Regular expressions are a more powerful alternative. Transcription can be represented as:

```
> open System.Text.RegularExpressions;;

> let transcribe dna =
    (new Regex("T")).Replace(dna, "U");;
val transcribe : string -> string
```

For example:

```
> transcribe "ACGTTGCAACGTTGCAACGTTGCA";;
val it : string = "ACGUUGCAACGUUGCAACGUUGCA"
```

Although regular expressions are overkill for this trivial example, they can be very useful for more sophisticated analyses.

6.3.2 Word frequency

This section details a simple program that uses a regular expression to separate words in order to accumulate the distribution of word frequencies in a given file. The implementation includes a general purpose counter that uses a Map to accumulate the frequency of each word.

Regular expressions can be used to identify words in strings, as sequences of alphabet characters. The following regular expression matches any sequence of one or more non-alphabet characters, i.e. word separators:

```
> open System.Text.RegularExpressions;;
> let not_alpha = new Regex(@"[^a-zA-z]+");;
val not_alpha : Regex
```

The following function uses the |> operator to compose a sequence of operations that, ultimately, present the words and their frequencies from the given file in descending order by number of occurrences:

```
> let freqs file =
    let a =
      lines_of_file file
      |> Seq.map not_alpha.Split
      |> Seq.concat
      |> Seq.filter (( <> ) "")
      |> Seq.map String.lowercase
      |> Seq.countBy (fun s -> s)
      |> Seq.to_array
    Array.sort (fun (_, n) (_, m) -> -compare n m) a
    a;;
val freqs : string -> (string * int) array
```

The lines in the given file are first obtained as a seq<string> using the lines_of_file function (defined in section 5.3). The Split member of the not_alpha regular expression is used to split each line into its constituent words. The sequence of sequences of words on each line are concatenated together to obtain the sequence of words in the file. Empty words are filtered out. The remaining words are converted into lowercase. Then the Seq.countBy function is used to count the number of occurrences of the key obtained by applying the given function (in this case the identity function) to each element. This gives the frequencies of the words in the file in an unspecified order. All of these operations are performed at once on each element when the laziness is forced into action by the Seq.to_array function.

The resulting array is then sorted into reverse order by the second elements of each of the pairs in the array. This gives the word frequencies with the most frequent words listed first.

For example, this freqs function takes only 2.9s to compute the word frequencies in the King James bible:

```
> time freqs @"C:\bible13.txt";;
```

```
Took 2911ms
val it : (string * int) array
  = [|("the", 64034); ("and", 51744); ("of", 34688);
      ("to", 13638); ("that", 12922); ("in", 12693);
      ("he", 10420); ("shall", 9838); ("unto", 8997);
      ("for", 8994); ("i", 8854); ("his", 8473); ...|]
```

Regular expressions have a wide variety of uses in scientific computing, ranging from simple string dissection to performing sophisticated computations over DNA sequences.

6.4   LIST RELATED

In this section, we shall examine a variety of functions that act upon lists. Consequently, we shall assume that the namespace of the List module has been opened:

```
> open List;;
```

These functions are often polymorphic and typically make use of either recursion and pattern matching or higher-order functions. These concepts can all be very useful but are rarely seen in current scientific programs.

6.4.1   count

The ability to count the number of elements in a list for which a predicate returns true is sometimes useful. A function to perform this task may be written most simply in terms of filter and length:

```
> let count f list =
    length (filter f list);;
val count : ('a -> bool) -> 'a list -> int
```

For example, the following counts the number of elements that are exactly divisible by three (0, 3, 6 and 9):

```
> count (fun x -> x % 3 = 0) [0 .. 9];;
val it : int = 4
```

However, the subexpression filter f list in this definition of the count function is generating an intermediate list unnecessarily. Consequently, this implementation of the count function can be deforested as described in section 8.4.4.1.

A faster count function might exploit the laziness of the Seq.filter function:

```
> let count f list =
    Seq.length (Seq.filter f list);;
val count : ('a -> bool) -> seq<'a> -> int
```

Note that this function is generic over the container type: it can be applied to arrays, sets and so on as well as lists.

Using the more specific `List.fold_left` function will be faster still because it avoids the overhead of laziness:

```
> let count f xs =
    fold_left (fun n x -> if f x then n+1 else n) 0 xs;;
val count : ('a -> bool) -> 'a list -> int
```

The `count` function can also be written using pattern matching:

```
> let rec count_aux n f = function
    | [] -> n
    | h::t when f h -> count_aux (n+1) f t
    | _::t -> count_aux n f t;;
val count_aux : int -> ('a -> bool) -> 'a list -> int
> let count list = count_aux 0 list;;
val count : ('a -> bool) -> 'a list -> int
```

However, this is more verbose than the fold-based implementation.

6.4.2 positions

The ability to prepend elements to lists indefinitely makes them the ideal data structure for many operations where the length of the output cannot be determined easily. Let us examine a function which composes an arbitrary length list as the result.

A function `positions` that returns the positions of elements satisying a predicate function `f` may be written in many different ways. For example:

```
> let positions f list =
    let cons_if (p, ps) h =
      p+1, if f h then p::ps else ps
    let _, ps = fold_left cons_if (0, []) list
    rev ps;;
val positions : ('a -> bool) -> 'a list -> int list
```

The nested auxiliary function `cons_if` accumulates the current index p and the list ps of positions satisfying the predicate function f. The positions function folds the `cons_if` function over the list, starting with an accumulator containing the index zero and the empty list. The final index is ignored and the `positions` function returns the reverse of the list of positions ps as it has been accumulated in reverse order (the first element to satisfy the predicate was prepended onto the empty list in the initial accumulator).

Like `count`, the `positions` function is useful for general purpose list dissection.

6.4.3 fold_to

Consider a higher-order function `fold_to` that folds partway through a list, returning the accumulator and the remaining tail. This function may be written:

```
> let rec fold_to i f accu list =
    match i, list with
    | 0, t -> accu, t
    | i, h::t -> fold_to (i-1) f (f accu h) t
    | _, [] -> accu, [];;
val fold_to :
  int -> ('a -> 'b -> 'a) -> 'a -> 'b list ->
    'a * 'b list
```

Note that this function returns an empty tail list if the index was out of bounds.

6.4.4 insert

Consider a function to insert an element at a given index in a list. This can be written elegantly in terms of the fold_to function, using the built-in List.rev_append function:

```
> let insert x i list =
    let rfront, back =
      fold_to i (fun t h -> h::t) [] list
    rev_append rfront (x::back);;
val insert : 'a -> int -> 'a list -> 'a list
```

For example, inserting 100 into the list $[0\ldots9]$ such that it appears at index 5:

```
> insert 100 5 [0 .. 9];;
val it : int list = [0; 1; 2; 3; 4; 100; 5; 6; 7; 8; 9]
```

This function is useful when performance is not a problem but the $T(i)$ complexity of this function renders it unsuitable for repeated insertions at random positions. If this functionality is required to be efficient then either the insertions should be amortized or the list should be replaced with a more suitable data structure.

6.4.5 chop

Consider a function to chop a list into two lists at a given index i, returning the two halves of the input list which we shall refer to as the *front* and the *back* lists. As this function terminates in the middle of the list, it is most easily written using pattern matching rather than in terms of the higher-order functions provided in the List module.

The chop function may be written succinct in terms of the fold_to function, accumulating the front list in reverse:

```
> let chop i list =
    let rfront, back =
      fold_to i (fun t h -> h::t) [] list
    rev rfront, back;;
val chop : int -> 'a list -> 'a list * 'a list
```

For example, the `chop` function may be used to split the list $[0 \ldots 10]$ into the lists $[0 \ldots 4]$ and $[5 \ldots 10]$:

```
> chop 5 [0 .. 10];;
val it : int list * int list =
  ([0; 1; 2; 3; 4], [5; 6; 7; 8; 9; 10])
```

Writing the `chop` function in terms of the naturally tail recursive `fold_to` function encouraged us to write a tail recursive `chop` function. Factoring out tail recursive higher-order functions like `fold_to` is a good way to improve clarity in order to keep many functions tail recursive.

This `chop` function can be used as a basis for more sophisticated list processing functions.

6.4.6 `dice`

Consider a function called `dice` that splits a list containing nm elements into n lists of m elements each. This function may be written in tail recursive form by accumulating the intermediate lists in reverse:

```
> let rec dice_aux accu m list =
    match fold_to m (fun t h -> h::t) [] list with
    | rfront, [] -> rev_map rev (rfront :: accu)
    | rfront, back -> dice_aux (rfront :: accu) m back;;
val dice_aux :
  'a list list -> int -> 'a list -> 'a list list
```

The `dice` function may then be written in terms of the `dice_aux` function by applying the empty list as the accumulator:

```
> let dice m list =
    dice_aux [] m list;;
val dice : int -> 'a list -> 'a list list
```

For example, the `dice` function may be used to dice the list $[1 \ldots 9]$ into 3 lists containing 3 elements each:

```
> dice 3 [1 .. 9];;
val it : int list list =
  [[1; 2; 3]; [4; 5; 6]; [7; 8; 9]]
```

Note that the action performed by dice can be reversed using the `flatten` function in the `List` module:

```
> flatten [[1; 2; 3]; [4; 5; 6]; [7; 8; 9]];;
val it : int list = [1; 2; 3; 4; 5; 6; 7; 8; 9]
```

The `dice` function could be used, for example, to convert a stream of numbers into 3D vectors represented by lists containing three elements.

6.4.7 `apply_at`

The ability to alter the ith element of a list using a given function is sometimes useful. As the ith element of a list may be reached by traversing the previous i elements, this task can be done in $\Theta(i)$ time complexity. A function to perform this task may be written in terms of the `fold_to` function (described in section 6.4.3) by replacing the head of the back list before appending the front list in reverse order using the `rev_append` function:

```
> let apply_at f i list =
    match fold_to i (fun t h -> h::t) [] list with
    | rfront, h::back -> rev_append rfront (f h::back)
    | _, [] -> invalid_arg "apply_at";;
val apply_at : ('a -> 'a) -> int -> 'a list -> 'a list
```

For example, the following replaces the 6th element (at the index 5) of the given list with its previous value plus twenty:

```
> apply_at (( + ) 20) 5 [0 .. 9];;
val it : int list = [0; 1; 2; 3; 4; 25; 6; 7; 8; 9]
```

Provided performance is unimportant, functions like this apply_at function can be very useful when manipulating lists.

6.4.8 `sub`

Another function found in the `Array` module but not in the `List` module is the `sub` function. This function extracts a subset of consecutive elements of length `len` starting at index `i`. A tail-recursive equivalent for lists may be written:

```
> let sub i len list =
    let (), after_i = fold_to i (fun () _ -> ()) () list
    fst (chop len after_i);;
val sub : int -> int -> 'a list -> 'a list
```

This implementation takes the back list after chopping at `i` and then chops this list at `len`, giving the result as the front list (extracted using the `fst` function).

For example, the 4-element sublist from $i = 3$ of the list $[0\dots9]$ is the list $[3\dots6]$:

```
> sub 3 4 [0 .. 9];;
val it : int list = [3; 4; 5; 6]
```

Just as `Array.sub` can be useful, so this `sub` function can come in handy in many different circumstances. However, the asymptotic complexity of this list sub function is worse than for arrays because this function must traverse the first `i + len` elements.

6.4.9 `extract`

A function similar to the `apply_at` function (described in section 6.4.7) but which *extracts* the ith element of a list, giving a 2-tuple containing the element and a list without that element, can also be useful. As for `apply_at`, the `extract` function may be written in terms of the `fold_to` function:

```
> let extract i list =
    match fold_to i (fun t h -> h::t) [] list with
    | rfront, h::back -> h, rev_append rfront back
    | _, [] -> invalid_arg "extract";;
val extract : int -> 'a list -> 'a * 'a list
```

For example, extracting the element with index five from the list $[0 \ldots 9]$ gives the element 5 and the list $[0 \ldots 4, 6 \ldots 9]$:

```
> extract 5 [0 .. 9];;
val it : int * int list =
  (5, [0; 1; 2; 3; 4; 6; 7; 8; 9])
```

This function has many uses, such as randomizing the order of elements in lists.

6.4.10 `shuffle`

This function can be used to randomize the order of the elements in a list, by repeatedly extracting randomly chosen elements to build up a new list:

```
> let rand = new System.Random;;
val rand : System.Random
> let rec shuffle_aux n accu = function
    | [] -> accu
    | t ->
        let h, t = extract (Random.int n) t
        shuffle_aux (n-1) (h :: accu) t;;
val shuffle_aux : int -> 'a list -> 'a list -> 'a list
```

This auxiliary function `shuffle_aux` counts down the length and accumulates the resulting list, so the `shuffle` function can be written by applying the precomputed list length and the empty list:

```
> let shuffle list =
    shuffle_aux (length list) [] list;;
val shuffle : 'a list -> 'a list
```

For example, applying the `shuffle` function to the list $\{0 \ldots 9\}$ gives a permutation containing the elements $0 \ldots 9$ in a random order:

```
> randomize [0; 1; 2; 3; 4; 5; 6; 7; 8; 9];;
val it : int list = [6; 9; 8; 5; 1; 0; 3; 2; 7; 4]
```

This function is useful in many situations. For example, the programs used to measure the performance of various algorithms presented in this book used this `shuffle` function to evaluate the necessary tests in a random order, to reduce systematic effects of garbage collection. However, the slow random access to lists renders the asymptotic complexity of this function $O(n^2)$. As we shall see in the next section, arrays can provide an $O(n)$ shuffle.

6.4.11 `transpose`

A function to transpose a rectangular list of lists is easily written in terms of pattern matching:

```
> let rec transpose = function
    | (_ :: _) :: _ as xss ->
        map hd xss :: transpose (map tl xss)
    | _ -> [];;
val transpose : 'a list list -> 'a list list
```

This function maps the hd function over the list of lists to obtain the first row of the transpose and prepends this onto the transpose of the remaining rows obtained by mapping the tl function over the list of lists. Unlike the Array2 module, the representation of a matrix as a list of lists does not restrict the number of columns in each row to a constant and, consequently, invalid data can be given to functions that expect rectangular input, such as this transpose function. In this case, improperly-shaped input will cause the hd function to raise an exception when it is applied to an empty list.

For example, the transpose of the matrix:

$$\begin{pmatrix} 1 & 2 & 3 \\ 4 & 5 & 6 \\ 7 & 8 & 9 \end{pmatrix}^T = \begin{pmatrix} 1 & 4 & 7 \\ 2 & 5 & 8 \\ 3 & 6 & 9 \end{pmatrix}$$

```
> transpose [[1; 2; 3]; [4; 5; 6]; [7; 8; 9]];;
val it : int list list =
  [[1; 4; 7]; [2; 5; 8]; [3; 6; 9]]
```

Many more sophisticated functions also act upon lists of lists, including combinatoric functions.

6.4.12 `combinations`

A function that computes the combinations that arise from composing lists with one element taken from each list in a given list of lists may be written:

```
> let rec combinations = function
    | [] -> [[]]
    | hs :: tss ->
        [ for h in hs
            for ts in combinations tss ->
              h :: ts ];;
val combinations : #seq<'a> list -> 'a list list
```

This `combinations` function prepends each head element onto each tail combination. Note that the use of comprehensions resulted in a generalization: the input to the `combinations` function is actually a list of sequences rather than a list of lists, i.e. the type `#seq<'a> list`.

For example, the following lists all combinations taken from the sets $(1, 2, 3)$, $(4, 5, 6)$ and $(7, 8, 9)$:

```
> combinations [[1; 2; 3]; [4; 5; 6]; [7; 8; 9]];;
val it : int list list =
  [[1; 4; 7]; [1; 4; 8]; [1; 4; 9]; [1; 5; 7];
   [1; 5; 8]; [1; 5; 9]; [1; 6; 7]; [1; 6; 8];
   [1; 6; 9]; [2; 4; 7]; [2; 4; 8]; [2; 4; 9];
   [2; 5; 7]; [2; 5; 8]; [2; 5; 9]; [2; 6; 7];
   [2; 6; 8]; [2; 6; 9]; [3; 4; 7]; [3; 4; 8];
   [3; 4; 9]; [3; 5; 7]; [3; 5; 8]; [3; 5; 9];
   [3; 6; 7]; [3; 6; 8]; [3; 6; 9]]
```

Combinations are one of the two most important combinatoric functions. The other is permutations.

6.4.13 `distribute`

The ability to compute all permutations of a list is sometimes useful. Permutations may be computed using a simple recurrence relation, by inserting the head of a list into all positions of the permutations of the tail of the list. Thus, a function to permute a list is most easily written in terms of a function which inserts the given element into the given n-element list at all $n + 1$ possible positions. This function is called `distribute`:

```
> let rec distribute e = function
    | [] -> [[e]]
    | x :: xs' as xs ->
        (e :: xs) ::
          [ for xs in distribute e xs' ->
              x :: xs ];;
val distribute : 'a -> 'a list -> 'a list list
```

This `distribute` function operates by prepending an answer, the element e prepended onto the given list l, onto the head of the given list prepended onto each of the distributions of the element e over the tail t of the given list.

For example, the following inserts the element 3 at each of the three possible positions in the list `[1; 2]`:

```
> distribute 3 [1; 2];;
val it : int list list =
  [[3; 1; 2]; [1; 3; 2]; [1; 2; 3]]
```

Permutations may be computed using this `distribute` function.

6.4.14 `permute`

A function to permute a given list may be written in terms of the `distribute` function:

```
> let rec permute = function
    | e :: t -> flatten (map (distribute e) (permute t))
    | [] -> [[]];;
val permute : 'a list -> 'a list list
```

This `permute` function then operates by distributing the head of the given list over the permutations of the tail.

For example, there are $3! = 6$ permutations of three values:

```
> permute [1; 2; 3];;
val it : int list list =
  [[1; 2; 3]; [2; 1; 3]; [2; 3; 1]; [1; 3; 2];
   [3; 1; 2]; [3; 2; 1]]
```

The `permute` function has many uses, including the combinatorial optimization of small problems. However, the permute function has $O(n \times n!)$ complexity. For $n = 10$, this function takes over 16s. Consequently, this function is not suitable for large combinatorial problems and, in fact, such problems can be very difficult or practically impossible to solve. Combinatorial optimization will be discussed in one of the complete examples in chapter 12.

6.4.15 Power set

The power set of a set A is the set of all subsets of A. This may be computed using a simple function:

```
> let rec powerset = function
    | [] -> [[]]
    | h::t ->
        [ for t in powerset t
            for t in [t; h::t] ->
              t ];;
val powerset : 'a list -> 'a list list
```

For example, the subsets of the set $\{1, 2, 3\}$ are:

```
> powerset [1;2;3];;
val it : int list list =
  [[]; [1]; [2]; [1; 2]; [3]; [1; 3]; [2; 3]; [1; 2; 3]]
```

This is an elegant use of the list comprehension syntax offered by F#.

6.5   ARRAY RELATED

Although arrays are currently overused in scientific computing, they are well suited to certain situations. Specifically, situations where memory is tight or where random access is a performance bottleneck.

6.5.1   `rotate`

The ability to rotate the elements of an array can sometimes be of use. This can be achieved by creating a new array, the elements of which are given by looking up the elements with rotated indices in the given array:

```
> let rotate i a =
    let n = Array.length a
    [|for k in 0 .. n - 1 ->
        let k = (k + i) % n
        a.[if k < 0 then n + k else k]|];;
val rotate : int -> 'a array -> 'a array
```

This function creates an array with the elements of `a` rotated left by `i`. For example, rotating two places to the left:

```
> rotate 2 [|0; 1; 2; 3; 4; 5; 6; 7; 8; 9|];;
val it : int array = [|2; 3; 4; 5; 6; 7; 8; 9; 0; 1|]
```

Rotating right can be achieved by specifying a negative value for `i`. For example, rotating right three places:

```
> rotate (-3) [|0; 1; 2; 3; 4; 5; 6; 7; 8; 9|];;
val it : int array = [|7; 8; 9; 0; 1; 2; 3; 4; 5; 6|]
```

Considering this function alone, the performance can be improved significantly by rotating the array elements in-place, by swapping pairs of elements. This can be regarded as a deforesting optimization (see section 8.4.4.1). However, the more elegant approach presented here can be refactored in the case of many subsequent rotations (and other, similar operations) such that no intermediate arrays need be created.

6.5.2   `swap`

A function to swap the ith and j^{th} elements of an array may be written:

```
> let swap (a : 'a array) i j =
    let t = a.[i]
    a.[i] <- a.[j]
    a.[j] <- t;;
val swap : 'a array -> int -> int -> unit
```

Note that this function swaps the elements in-place and, consequently, returns the value `()` of type unit.

This function can be used as the basis of many algorithms that permute the contents of an array, such as sorting algorithms.

6.5.3 except

A useful function in the context of array manipulation takes integers $i \in \{0 \ldots n-1\}$ and $j \in \{0 \ldots n-2\}$ and returns the jth index that is not i:

```
> let except i j =
    if j < i then j else j + 1;;
val except : int -> int -> int
```

The except function can be used to shuffle array elements.

6.5.4 shuffle

The order of the elements in an array can be randomised in-place by swapping each element with with a randomly-chosen element. A function to perform this operation as a side effect may be written in terms of the swap function and the Random module presented in section 9.4:

```
> let shuffle a =
    let n = Array.length a
    for i = 0 to n - 1 do
      swap a i (Random.int n);;
val shuffle : 'a array -> unit
```

Note that we are careful to choose any element at random such that an element may be swapped with itself. Although it is tempting to swap with other elements, this would lead to determinism, e.g. when shuffling a two-element array.

For example:

```
> let a = [|1 .. 9|];;
val a : int array
> shuffle a;;
val it : unit = ()
> a;;
val it : int array = [|5; 4; 7; 3; 9; 1; 2; 6; 8|]
```

This function can be used in a wide variety of random algorithms and is particularly useful when performance is important.

6.6 HIGHER-ORDER FUNCTIONS

As we have already hinted, aggressively factoring higher-order functions can greatly reduce code size and sometimes even lead to a better understanding of the problem. In this section, we shall consider various different forms of higher-order functions which can be productively used to aid brevity and, therefore, clarity.

6.6.1 Tuple related

Functions to perform operations such as map over tuples of a particular arity are also useful. For example, the following implements some useful functions over 2-tuples:

```
> let iter_2 f (a, b) =
    f a
    f b;;
val iter_2 : ('a -> unit) -> 'a * 'a -> unit
> let map_2 f (a, b) =
    f a, f b;;
val map_2 : ('a -> 'b) -> 'a * 'a -> 'b * 'b
> let fold_left_2 f accu (a, b) =
    f (f accu a) b;;
val fold_left_2 :
  ('a -> 'b -> 'a) -> 'a -> 'b * 'b -> 'a
```

Such functions can be used to reduce code size in many cases.

6.6.2 Generalized products

The vector dot product is a specialized form of inner product. The inner and outer products may, therefore, be productively written as higher-order functions which can then be used as a basis for more specialized products, such as the dot product.

The inner product is most easily written in terms of a fold_left2 function. In the interests of generality, we shall begin by defining a fold_left2 function for the Seq type:

```
> let fold_left2 f accu xs ys =
    Seq.zip xs ys
    |> Seq.fold (fun accu (x, y) -> f accu x y) accu;;
val fold_left2 :
  ('a -> 'b -> 'c -> 'a) -> 'a -> #seq<'b> ->
    #seq<'c> -> 'a
```

An inner product that is generalized over types, functions and even container type may then be written in terms of this fold_left2 function:

```
> let inner base f xs ys g =
    fold_left2 (fun accu x y -> g accu (f x y))
      base xs ys;;
val inner :
  'a -> ('b -> 'c -> 'd) -> #seq<'b> -> #seq<'c> ->
    ('a -> 'd -> 'a) -> 'a
```

For example, the same inner function may be used to compute the dot product of a pair of int lists:

```
> inner 0 ( * ) [1; 2; 3] [2; 3; 4] ( + );;
```

```
val it : int = 20
```

float lists:

```
> inner 0.0 ( * ) [1.0; 2.0; 3.0] [2.0; 3.0; 4.0]
    ( + );;
val it : float = 20.0
```

and even float arrays:

```
> inner 0.0 ( * ) [|1.0; 2.0; 3.0|] [|2.0; 3.0; 4.0|]
    ( + );;
val it : float = 20.0
```

Using the Seq type, even the outer product can be generalized over data structure:

```
> let outer f xs ys =
    seq { for x in xs ->
            seq { for y in ys ->
                    f x y } };;
val outer :
  ('a -> 'b -> 'c) -> #seq<'a> -> #seq<'b> ->
    seq<seq<'c>>
```

The outer product of two vectors is a matrix:

$$(1,2,3) \otimes (2,3,4) = \begin{pmatrix} 2 & 3 & 4 \\ 4 & 6 & 8 \\ 6 & 9 & 12 \end{pmatrix}$$

The generalized outer function can be used to compute this outer product on many container types including float lists:

```
> outer ( * ) [1.0; 2.0; 3.0] [2.0; 3.0; 4.0];;
val it : seq<seq<float>>
  = seq [seq [2.0; 4.0; 6.0]; seq [3.0; 6.0; 9.0];
         seq [4.0; 8.0; 12.0]]
```

Aggressive factoring of higher-order functions can be very useful in the context of numerical computation.

CHAPTER 7

VISUALIZATION

The ability to visualize problems and data can be of great use when trying to understand difficult concepts and, hence, can be of great use to scientists. Perhaps the most obvious application of visualization in science is in the study of molecules, particularly biological molecules. However, a great many other problems can also be elucidated through the use of visualization, particularly real-time and interactive graphics which can go well beyond the capabilities of the previous generation of static, made-for-print graphics.

This chapter introduces two tools that are essential for the development of sophisticated graphical applications on the Windows platform:

- *Windows Forms* for writing graphical user interfaces (**GUIs**).

- *Managed DirectX* for generating high-performance, real-time and interactive graphics.

These tools are then used to develop a library that allows easy-to-use graphical applications to be spawned seamlessly from the F# interactive mode, and an example stand-alone graphical application for visualizing scientific data.

Several of the complete example programs presented in chapter 12 use the scene graph library developed in this chapter.

F# for Scientists. By Jon Harrop
Copyright © 2008 John Wiley & Sons, Inc.

7.1 WINDOWS FORMS

A great deal is expected of modern computer programs. They should provide a graphical user interface that presents the necessary information to the user in a clear and comprehensible way.

Fortunately, Windows Forms provides an easy way to create graphical user interfaces. Using Visual Studio, GUIs can be designed using drag and drop and the generated C# code can be edited to provide the necessary functions, responding to events such as button clicks in order to provide the required interface. In many cases, only a simple GUI is required and, in such cases, this can be achieved without the benefit of a GUI designer.

This section explains how simple GUIs can be created by writing only F# code. In particular, how GUIs can be spawned from the F# interactive mode in order to test and tweak them.

7.1.1 Forms

The most fundamental construct in Windows Forms is the form. Definitions provided by .NET that are related to Windows Forms are in the namespace:

```
> open System.Windows.Forms;;
```

A blank form can be created by a single line of code, calling the constructor of the .NET Form class and setting some properties of the form to make it visible and give it a title:

```
> let form = new Form(Visible=true, Text="First form");;
val form : Form
```

The resulting form is shown in figure 7.1.

Setting the TopMost property ensures that the form stays as the top-most window:

```
> form.TopMost <- true;;
val it : unit = ()
```

This is useful when developing applications from the F# interactive mode, as the interactive session can be used to develop the form that is shown on top.

Graphical user interfaces inside forms are composed of *controls*, such as labels, text boxes and even web browsers. The layout of controls on a form is often a function of the size of the form. The size of a form is accessible in two different ways. The total size of the form including window decorations (the title bar, scroll bars and so on) can be obtained using:

```
> form.Size;;
val it : Size = {Width=300, Height=300} ...
```

The area available for controls is often a more important quantity and is called the client rectangle. The size of the client rectangle is slightly smaller than the size of the entire form:

Figure 7.1   A blank Windows form.

```
> form.ClientSize;;
val it : Size = {Width=292, Height=260} ...
```

In order to make a useful GUI, controls must be added to the form.

7.1.2   Controls

Creating and adding controls to forms is also easy. A control is created by calling the appropriate constructor, for example a button:

```
> let button = new Button(Text="A button");;
val button : Button
```

The control can then be added to the form using the form's `Controls.Add` member:

```
> form.Controls.Add(button);;
val it : unit = ()
```

The resulting form is shown in figure 7.2.

7.1.3   Events

For a suite of controls laid out on a form to be useful, the controls must be connected to functions that are executed when events occur, such as the clicking of a button. This is achieved by adding functions as event handlers.

The button in the previous example does nothing when it is clicked. To make the button click perform an action, a function called an event handler must be added to

Figure 7.2 A form with a single control, a button.

the `button.Click` event. In F#, the event handler can be an anonymous function. For example, to close the form when the button is clicked:

```
> button.Click.Add(fun _ -> form.Close());;
val it : unit = ()
```

Constructing controls and registering event handlers in this style allows GUIs to be composed with relative ease and, in particular, to be developed from an F# interactive session.

7.1.4 Bitmaps

Before delving into fully-fledged DirectX programming, it is worth taking a peek at the functionality provided by simple bitmaps rendered using Windows Forms. Bitmaps can be used to render a variety of simple, raster-based images. In this section we shall outline a simple programmatic bitmap renderer before using it to render a cellular automaton.

Many relevant definition are in the following namespaces:

```
> open System;;
> open Drawing;;
> open Windows.Forms;;
```

In this case, the bitmap will have 24 bits-per-pixel (bpp):

```
> let format = Imaging.PixelFormat.Format24bppRgb;;
val format : Imaging.PixelFormat
```

A bitmap can be created to cover the client area of a form and filled using the given function f:

```
> let bitmap_of f (r : Rectangle) =
    let bitmap = new Bitmap(r.Width, r.Height, format)
    f bitmap
    bitmap;;
val bitmap_of : (Bitmap -> unit) -> Rectangle -> Bitmap
```

An event handler can used used to replace the bitmap when the size of the form changes:

```
> let resize f (b : Bitmap ref) (w : #Form) _ =
    b := bitmap_of f w.ClientRectangle
    w.Invalidate();;
val resize :
  (Bitmap -> unit) -> Bitmap ref -> #Form -> 'b -> unit
```

An event handler can be used to draw the bitmap into the window:

```
> let paint (b : Bitmap ref) (v : #Form)
      (e : PaintEventArgs) =
    let r = e.ClipRectangle
    e.Graphics.DrawImage(!b, r, r, GraphicsUnit.Pixel);;
val paint :
  Bitmap ref -> #Form -> PaintEventArgs -> unit
```

The following function creates a form, a bitmap and registers handlers for resizing, painting and key press (to close the form if the escape key is pressed):

```
> let make_raster f =
    let form = new Form(Visible=true)
    let bitmap = ref (bitmap_of f form.ClientRectangle)
    form.Resize.Add(resize f bitmap form)
    form.Paint.Add(paint bitmap form)
    form.KeyDown.Add(fun e ->
      if e.KeyCode = Keys.Escape then form.Close())
    form;;
val make_raster : string -> (Bitmap -> unit) -> Form
```

The make_raster function can be used as the basis of many programs that require simple bitmap output. Rendering is left entirely up to the user using the function f to draw into the bitmap as necessary.

7.1.5   Example: Cellular automata

One application ideally suited to rendering into bitmaps is cellular automata. In Stephen Wolfram's seminal work "A New Kind of Science", he studies a simple but remarkably unpredictable 1D cellular automaton called rule 30 where each generation is a row of cells that are in one of two states.

The two states of a cell may be represented by a variant type:

```
> type cell = A | B;;
```

A rule computes the state of a cell in the next generation given the state of the cell and its two neighbours in the current generation. Rule 30 may be written:

```
> let rule30 a b c =
    match a, b, c with
    | B, B, _ | B, A, B | A, A, A -> A
    | _ -> B;;
val rule30 : cell -> cell -> cell -> cell
```

A generation may be evolved into the next generation by mapping a rule over triples of neighbouring cells in the current generation, assuming cells outside to be in the state A:

```
> let rec evolve_aux rule = function
    | a::(b::c::_ as t) ->
        rule a b c :: evolve_aux rule t
    | [a; b] -> [rule a b A; rule b A A]
    | [a] -> [rule A a A; rule a A A]
    | [] -> [rule A A A];;
val evolve_aux :
  (cell -> cell -> cell -> 'a) -> cell list -> 'a list
```

The auxiliary function evolve_aux that computes the next generation is finished by prepending the left-most cell and padding the previous generation with a cell in state A before passing it to the evolve_aux function:

```
> let evolve rule list =
    evolve_aux rule (A :: A :: list);;
val evolve :
  (cell -> cell -> cell -> 'a) -> cell list -> 'a list
```

Cells in state A will be drawn in white and cells in state B will be drawn in black:

```
> let color_of_cell = function
    | A -> Color.White
    | B -> Color.Black;;
val color_of_cell : cell -> Color
```

A single cell is drawn by shifting the x coordinate to "grow" the generations from the center of the bitmap and filling the pixel if it is visible:

```
> let set (bitmap : Bitmap) y x c =
    let x = bitmap.Width / 2 - y + x
    if 0 <= x && x < bitmap.Width then
      bitmap.SetPixel(x, y, color_of_cell c);;
val set : Bitmap -> int -> int -> cell -> unit
```

The whole bitmap is drawn by evolving a generation for each row and filling each cell in each row:

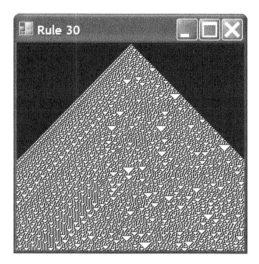

Figure 7.3   A thousand generations of the rule 30 cellular automaton.

```
> let draw rule bitmap =
    let aux gen y =
      List.iteri (set bitmap y) gen
      evolve rule gen
    ignore (Seq.fold aux [B] {0 .. bitmap.Height - 1});;
val draw :
  (cell -> cell -> cell -> cell) -> Bitmap -> unit
```

Finally, a form rendering this bitmap may be created using the make_raster function defined in the previous section:

```
> let form = make_raster (draw rule30);;
val form : Form
```

The resulting visualization is illustrated in figure 7.3.

7.1.6   Running an application

If the above code is compiled into an application and run the program will quit immediately and close the window. To get the desired behaviour, the main thread of the program should block until the form is closed. This is achieved by calling the Application.Run function.

If a program spawns a single window, or if there is a clear master window in the GUI design, a compiled application may block until a specific form is closed by calling:

```
Application.Run(form)
```

where form is the master window.

Alternatively, the `Application.Exit` function may be used to exit a program explicitly. Such a program can be run by supplying the value of type unit to the `Run` member, rather than a form:

```
Application.Run()
```

In many cases it is desirable to provide an F# program such that it can be executed either from an interactive session or compiled into a program and run. In such cases, the `#if` preprocessor directive can be used to select code depending upon the way in which the program is run by testing either `COMPILED` or `INTERACTIVE`:

```
#if COMPILED
Application.Run(form)
#endif
```

This allows identical code to be used both in the interactive mode and compiled to standalone executable.

More advanced applications require the ability run separate windows concurrently. This requires the use of threading and will be discussed in more detail later in this chapter, in the context of spawning visualizations from F# interactive sessions, in section 7.2.5.

7.2 MANAGED DIRECTX

The defacto standard for high-performance graphics on the Windows platform is *DirectX*. Microsoft provide a high-level interface to DirectX from .NET, known as *Managed DirectX*.

Despite being a high-level interface, programs using Managed DirectX must contain a significant amount of "boiler plate" code that is required to get anything working. This common code is presented here in the form of a reusable library written in F# that allows visualizations to be spawned from the F# interactive mode.

7.2.1 Handling DirectX devices

All programs that use DirectX must begin with declarations to include the appropriate directory:

```
> #I @"C:\WINDOWS\Microsoft.NET\DirectX for Managed
Code\1.0.2902.0";;
```

and import the relevant DLLs:

```
> #r @"Microsoft.DirectX.dll";;
> #r @"Microsoft.DirectX.Direct3D.dll";;
> #r @"Microsoft.DirectX.Direct3DX.dll";;
```

Graphics-related functions make heavy use of several namespaces, which can be productively opened to improve clarity:

```
> open System;;
> open Drawing;;
> open Windows.Forms;;
> open Microsoft.DirectX;;
> open Direct3D;;
```

The properties of a DirectX device are dictated by a set of *present parameters*[17]:

```
> let presentParams () =
    let p = new PresentParameters(Windowed=true)
    p.SwapEffect <- SwapEffect.Discard
    p.EnableAutoDepthStencil <- true
    p.AutoDepthStencilFormat <- DepthFormat.D24S8
    [| p |];;
val presentParams : unit -> PresentParameters array
```

A DirectX device with these present parameters can be created by calling the constructor of the `Device` class:

```
> let make_device (form : #Form) =
    let dtype = DeviceType.Hardware
    let flags = CreateFlags.HardwareVertexProcessing
    new Device(0, dtype, form, flags, presentParams());;
val make_device : #Form -> Device
```

As .NET libraries such as DirectX are written in an object oriented style, this style can be adopted in the F# code that interfaces with these libraries. In this case, a minimal `Viewer` class is used to provide a form with no background using the implicit-constructor form of class declaration described in section 2.4.2.2:

```
> type Viewer () =
    inherit Form()
    override form.OnPaintBackground _ = ();;
```

In effect, we shall be specializing the `Form` class in order to implement the functionality of a DirectX viewer. However, rather than adopting the C# style of writing one large class, we shall instead opt to develop several small functions that can be retrofitted onto this minimal `Viewer` class to provide the desired functionality.

A higher-order `paint` callback can be used to perform mundane tasks including calling the `render` function to render the scene, which is passed as an argument to `paint`:

```
> let paint (form : #Form) render (device : Device) _ =
    try
      device.TestCooperativeLevel()
      device.BeginScene()
      render device
```

---

[17]Note that the homonym "present" in the context of DirectX typically means "show" rather than "now".

```
        device.EndScene()
        device.Present()
        form.Invalidate()
    with
        | :?  DeviceLostException -> ()
        | :?  DeviceNotResetException ->
            device.Reset(presentParams());;
val paint :
  #Form -> (Device -> unit) -> Device -> 'b -> unit
```

Note the use of the :? construct to handle exceptions generated outside F#.

The complexity of this paint function is due to the fact that managed DirectX requires us to handle situations where the DirectX device is left corrupted by another application (a *device reset*) or lost entirely.

A Viewer object and DirectX device can be created and initialized using a higher-order make_viewer function:

```
> let make_viewer title render =
    let form = new Viewer(Text=title, Visible=true)
    form.MinimumSize <- form.Size
    let device = make_device form
    form.Paint.Add(paint form render device)
    form.Resize.Add(fun _ -> form.Invalidate())
    form.KeyDown.Add(fun e ->
      if e.KeyCode = Keys.Escape then form.Close())
    form.Invalidate()
    form;;
val make_viewer : string -> (Device -> unit) -> Viewer
```

This function is careful to constrain the minimum size of the window because managed DirectX will crash if a device is shrunk too much.

By default, the contents of a window are redrawn when the window is enlarged but not when it is shrunk. In order to redraw the window whenever it is resized, the Resize event is made to invalidate the form. The KeyDown event is used to close the form if the escape key is pressed.

A minimal DirectX renderer that simply clears the display to a specified color can now be run from only 3 lines of code. The render function clears the *target buffer* of the device to a given color:

```
> let render (device : Device) =
    let target = ClearFlags.Target
    device.Clear(target, Color.Coral, 1.f, 0);;
val render : Device -> unit
```

The make_viewer function can then be used to create a blank form by providing the title and render function:

```
> let form = make_viewer "Blank" render;;
```

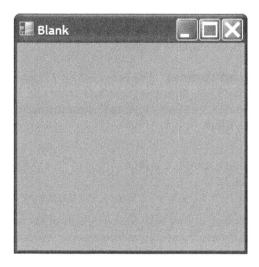

Figure 7.4   A DirectX viewer that clears the display to a single color (called "coral").

```
val form : Viewer
```

The resulting window is illustrated in figure 7.4.

7.2.2   Programmatic rendering

DirectX is designed for *programmatic rendering*, where programs explicitly invoke the DirectX API to get a scene rendered. This section describes how the DirectX API can be invoked directly from F# code. Programmatic rendering has the advantage that all aspects of rendering can be completely controlled and, consequently, programs can be tweaked to improve performance. Later in this chapter, we shall introduce a simpler way to visualize data using *declarative rendering* from an F# interactive session.

In the simplest case, programmatic rendering can be broken down into per-pixel buffers, vectors, vertices, projections, views and the rendering of primitives. Each of these topics will now be discussed before moving onto more sophisticated subjects.

7.2.2.1   Buffers   A DirectX device provides up to three different kinds of pixel buffer: the target buffer for pixel colors, the z-buffer to store depth values and the stencil buffer for per-pixel clipping. The paint event will almost always clear all three buffers and, consequently, it is useful to factor out a value representing all three buffers:

```
> let all_buffers =
    Enum.combine
      [ ClearFlags.Target;
        ClearFlags.ZBuffer;
        ClearFlags.Stencil ];;
```

```
val all_buffers : ClearFlags
```

This value will be used in future functions.

7.2.2.2 Vectors and vertices

The native 64-bit `float` type of F# is more accurate than the native 32-bit format of DirectX (`float32`). Trading performance for simplicity, the DirectX `Vector3` type can be constructed using an F# function that handles `float` values:

```
> let vec(x, y, z) =
    let f = Float32.of_float
    new Vector3(f x, f y, f z);;
val vec : float * float * float -> Vector3
```

In the interests of simplicity, we shall use a single `Vertex` type that encapsulates the coordinate, normal vectorz and color information for every vertex:

```
> type Vertex = CustomVertex.PositionNormalColored;;
```

An F#-friendly constructor for the `Vertex` type may be written in curried form:

```
> let vertex (c : Color) (nx, ny, nz) (x, y, z) =
    let c = c.ToArgb()
    let f = Float32.of_float
    new Vertex(f x, f y, f z, f nx, f ny, f nz, c);;
val vertex :
  Color -> (float * float * float) ->
    (float * float * float) -> Vertex
```

The order of the parameters to the curried `vertex` function have been chosen such that partial application of this function is as useful as possible. Specifically, whole objects are likely to share a single color, which can be partially applied first. Individial faces may be composed of several triangles sharing a single normal vector, which can be partially applied next. Finally, the vertex coordinate is applied to obtain a `Vertex` value.

The following function creates a `Vertex` from two coordinates:

```
> let vertex2 color (x, y) =
    vertex color (0.0, 0.0, -1.0) (x, y, 0.0);;
val vertex2 : Color -> float -> float -> Vertex
```

These functions will be used to compose programs that render geometry using DirectX.

7.2.2.3 Orthographic projection

An orthographic projection can be used to visualize 2D scenes. The following `show2d` combinator initializes an orthographic projection before calling the given `render` function:

```
> let show2d render (device : Device) =
    let near, far = -1.0f, 1.0f
```

```
    let vp = device.Viewport
    let w, h = float32 vp.Width, float32 vp.Height
    device.Transform.Projection <-
      Matrix.OrthoLH(w, h, near, far)
    device.Clear(ClearFlags.Target, Color.White, far, 0)
    device.RenderState.Lighting <- false
    device.RenderState.CullMode <- Cull.None
    device.RenderState.ZBufferEnable <- false
    render device;;
val show2d : (Device -> unit) -> Device -> unit
```

This combinator sets the near and far clipping planes to $z = -1$ and $z = 1$, respectively, in anticipation of vertex coordinates on the plane $z = 0$. The viewport is set to $\pm w/2$ and $\pm h/2$ where w and h are the width and height of the DirectX device (the whole of the client rectangle of the window), i.e. the coordinates in the scene are measured in pixels and the origin is the center of the window. The buffers are cleared and lighting and culling are disabled before the `render` function is called to render the current scene.

As DirectX is primarily used to render 3D scenes that often contain closed meshes of triangles, it provides the ability to neglect triangles that are back-facing as an optimization. A triangle is deemed to be back-facing or front-facing depending on whether its vertices are listed in clockwise or counter-clockwise order from the point of view of the camera. In the 2D case, this backface culling is not usually useful, so our `show2d` combinator disables this default behaviour.

7.2.2.4 *Perspective projection* Rendering graphics in 3D is only slightly more difficult than the 2D case but is often more useful. Setting up 3D visualization typically requires perspective projection and lighting. This section defines several functions that can be used to simplify the task of programming 3D graphics in F#.

The aspect ratio of the display must be determined in order to create a 3D perspective projection. This can be obtained via the viewport of the DirectX device using the following function:

```
> let aspect (device : Device) =
    let vp = device.Viewport
    float32 vp.Width / float32 vp.Height;;
val aspect : Device -> float32
```

A perspective projection is quantified by a field of view (**FOV**), the distances of near and far clipping planes from the camera and the positions of the camera, target (that the camera points at) and the up-direction of the camera:

```
> type perspective =
    {
      fov: float32;
      aspect: float32;
      near: float32;
```

```
      far: float32;
      camera: Vector3;
      target: Vector3;
      up: Vector3;
    };;
```

We shall use the following default values:

```
> let perspective_default =
    { fov = 0.8f; aspect = 0.0f;
      near = 0.01f; far = 100.0f;
      camera = vec(-1.0, 2.0, -4.0);
      target = vec(0.0, 0.0, 0.0);
      up = vec(0.0, 1.0, 0.0) };;
val perspective_default : perspective
```

The following function sets a perspective projection and view in the DirectX device:

```
> let perspective p (t : Transforms) =
    t.Projection <-
      Matrix.PerspectiveFovLH(p.fov, p.aspect,
                              p.near, p.far)
    t.View <-
      Matrix.LookAtLH(p.camera, p.target, p.up);;
val perspective : perspective -> Transforms -> unit
```

The `Projection` and `View` matrices of the DirectX device are setup by the `show3d` combinator before it calls the given `render` function:

```
> let show3d render (device : Device) =
    let p = { perspective_default with
                aspect = aspect device }
    perspective p device.Transform
    device.Clear(all_buffers, Color.White, p.far, 0)
    device.Lights.[0].Direction <- vec(1.0, -1.0, 2.0)
    device.Lights.[0].Enabled <- true
    render device;;
val show3d : (Device -> unit) -> Device -> unit
```

This function is suitable for displaying objects at the origin and of roughly unit radius. A single light is set. Note that culling defaults to backface culling, so triangles are only visible from one side.

In order to render a 2D or 3D scene it is necessary to provide the `show2d` or `show3d` functions with a `render` function that can render primitives onto the device.

7.2.2.5 *Rendering primitives*   Modern graphics hardware is capable of rendering a variety of primitives (see figure 7.5). All non-trivial objects, such as spheres and cylinders, must be approximated using these primitives.

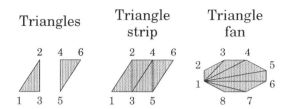

Figure 7.5   Abutting triangles can be amortised into triangle fans and strips to reduce the number of vertices required to describe a geometry.

Triangle strips and fans are used as more space efficient representations of abutting triangles, requiring $n+2$ vertices to describe n triangle rather than $3n$ vertices required when triangles are specified individually. Although this is a productive optimization, we shall restrict ourselves to the rendering of individual triangles in this chapter in the interests of simplicity.

Table 7.1   DirectX primitive drawing functions.

	Direct from vertex array	Indirected through index array
Ordinary arrays	DrawUserPrimitives	DrawIndexedUserPrimitives
Compiled arrays	DrawPrimitives	DrawIndexedPrimitives

Primitives can be drawn by calling one of several members of the `Device` class:

- `DrawUserPrimitives`

- `DrawIndexedUserPrimitives`

- `DrawPrimitives`

- `DrawIndexedPrimitives`

These four functions have different properties (see table 7.1).

The `DrawUserPrimitives` and `DrawIndexedUserPrimitives` functions render primitives from a vertex array specified as a native .NET (or F#) array. These functions are less efficient but easier to use.

The `DrawPrimitives` and `DrawIndexedPrimitives` functions render primitives from compiled vertex arrays. These functions are more efficient but more complicated and harder to use. Specifically, these functions require the vertex array to be in a compiled form and, in particular, the properties of the compiled form can be set such that the vertex data actually resides on the graphics card, dramatically increasing performance for static vertex data.

In the interests of simplicity, we shall use only the `DrawUserPrimitives` function in the remainder of this chapter.

The F# programming language can provide more compile-time assurances than the Managed DirectX API currently provides. Specifically, the vertex and index data passed to these functions is not statically typed. In order to catch errors at compile time, we shall wrap the `DrawUserPrimitives` function in a statically typed function to render individual triangles:

```
> let draw_triangles (device : Device)
      (vertex : Vertex array) =
    device.VertexFormat <- Vertex.Format
    let prim = PrimitiveType.TriangleList
    let n = Array.length vertex / 3
    device.DrawUserPrimitives(prim, n, vertex);;
val draw_triangles : Device -> Vertex array -> unit
```

This function expects a vertex array containing $3n$ vertices, 3 for each of n triangles.

A window visualizing a triangle can be spawned from the F# interactive mode with:

```
> let form =
    let render device =
      [| -100.0, -100.0; 0.0, 100.0; 100.0, -100.0 |]
      |> Array.map (vertex2 Color.BurlyWood)
      |> draw_triangles device
    show2d render
    |> make_viewer "Triangle";;
val form : Form
```

The result is illustrated in figure 7.6.

7.2.3 Rendering an icosahedron

The 12 vertices of an icosahedron are given by:

```
> let vertices =
    let r, y = sin(Math.PI / 3.0), cos(Math.PI / 3.0)
    let f g x = r * g(float x * Math.PI / 5.0)
    let aux n y i =
      vec(f sin (2*i + n), y, f cos (2*i + n)))
    Array.concat
      [ [|vec(0.0, 1.0, 0.0)|];
        Array.init 5 (aux 0 y);
        Array.init 5 (aux 1 (-y));
        [|vec(0.0, -1.0, 0.0)|] ];;
val vertices : Vector3 array
```

The 20 triangular faces of an icosahedron are given in terms of those vertices by:

```
> let faces =
```

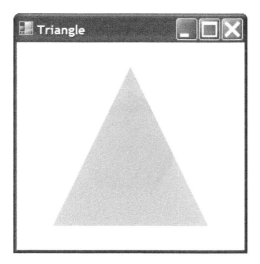

Figure 7.6   A triangle rendered programmatically and visualized using an orthographic projection.

```
    let r1 i = 1 + (1 + i) % 5
    let r2 i = 6 + (1 + i) % 5
    [ for i in 0 .. 4 -> 0, r1 i, r1(i + 1) ] @
    [ for i in 0 .. 4 -> r1(i + 1), r1 i, r2 i ] @
    [ for i in 0 .. 4 -> r1(i + 1), r2 i, r2(i + 1) ] @
    [ for i in 0 .. 4 -> r2(i + 1), r2 i, 11 ];;
val faces : (int * int * int) list
```

The following `tri_of_face` function converts the color and vertices of a triangular face from the icosahedron into a vertex array ready for rendering:

```
> let tri_of_face (c : Color) (v : Vector3 array)
      (i, j, k) =
    let i, j, k = v.[i], v.[j], v.[k]
    let n = Vector3.Normalize(i + j + k)
    let v r = new Vertex(r, n, c.ToArgb())
    [| v i; v j; v k |];;
val tri_of_face :
  Color -> Vector3 array -> int * int * int ->
    Vertex list
```

A complete vertex array for the icosahedron may then be created by concatenating the result of mapping the `tri_of_face` function over the faces of the icosahedron:

```
> let triangles =
    faces
    |> List.map (tri_of_face Color.BurlyWood vertices)
    |> Array.concat;;
```

Figure 7.7   A DirectX viewer that draws an icosahedron.

```
val triangle : Vertex array
```

This demo can be animated to make it more interesting. The simplest way to define an animation in a functional programming language is to make the scene a function of time. The elapsed time can be obtained from a running stopwatch:

```
> let timer = new System.Diagnostics.Stopwatch();;
val timer : Diagnostics.Stopwatch

> timer.Start();;
```

The rendering function simply sets the world transformation matrix to a rotation about the y-axis and then draws the triangles from the vertex array:

```
> let render_icosahedron (device : Device) =
    let t = float32 timer.ElapsedMilliseconds / 1e3f
    device.Transform.World <- Matrix.RotationY t
    draw_triangles device triangles;;
val render_icosahedron : Device -> unit
```

The make_viewer function and show3d combinator may then be used to visualize the icosahedron:

```
> let form =
    show3d render_icosahedron
    |> make_viewer "Icosahedron";;
val form : Viewer
```

The result is illustrated in figure 7.7.

As this example has demonstrated, animated 3D graphics can be visualized with little effort in F#. However, these examples used F# code to render a scene programmatically. Visualizations typically have so much in common that the only variable is the data itself. Consequently, declarative rendering is prolific in scientific computing.

7.2.4 Declarative rendering

A scene can be represented by a single value, a data structure, that conveys all of the necessary information about the positions and colors of triangles in the scene. This allows scenes to be visualized without any programming. In a functional language, the value representing a scene may contain functions and, in particular, may be a function of time to facilitate animation.

This section describes additional code that can be used to allow real-time, animated 2D and 3D graphics to be visualized as easily as possible from the F# interactive mode, by describing scenes as values. Although this is a comparatively small amount of library code, the functionality of the library actually exceeds the functionality provided by several expensive commercial packages.

The type of value used to represent a scene is known as a *scene graph*.

7.2.4.1 *Scene graph* In F#, a variant type can be used to represent a scene:

```
> type scene =
    | Triangles of (Vector3 * Vector3 * Vector3) list
    | Color of Color * scene
    | Transform of Matrix * scene
    | Group of scene list;;
```

In this case, there are four different kinds of scene graph node. The `Triangles` constructor represents a set of triangles as a list of 3-tuples of 3D vectors. The `Color` constructor allows the color of triangles given in the child scene to be overriden. The `Transform` constructor allows a matrix transformation (such as a rotation or scaling) to be applied to the child scene. Finally, the `Group` constructor allows separate scenes to be composed.

A simple scene containing a single triangle can now be described very succinctly in terms of the `scene` variant type:

```
> Triangles
    [ vec(-1.0, 0.0, 0.0),
      vec(0.0, 1.0, 0.0),
      vec(1.0, 0.0, 0.0) ];;
val it : scene = Triangles...
```

As we shall see, the ability to visualize a scene defined in this way is extremely useful, particularly in the context of scientific computing.

7.2.4.2 *Rendering a scene graph* In the interests of generality, we shall continue to use the `Vertex` type that includes position, normal and color information for every vertex. However, the `scene` type does not allow normal vectors and colors to be defined per vertex.

The following function converts a triangle from the scene graph representation into a vertex array suitable for rendering:

```
> let triangle (c : Color) (p0, p1, p2) =
    let p01, p02 = p1 - p0, p2 - p0
    let n = Vector3.Normalize(Vector3.Cross(p01, p02))
    Array.map (vertex c n) [|p0; p1; p2|];;
val triangle :
  Color -> Vector3 * Vector3 * Vector3 -> Vertex array
```

A Transform node of a scene can be rendered using a combinator that multiplies the current World transformation matrix by the transformation matrix m of the node, calls the given function k and restores the World matrix before returning:

```
> let transform (device : Device) color m k scene =
    let world = device.Transform.World
    device.Transform.World <- Matrix.Multiply(world, m)
    try k device color scene finally
    device.Transform.World <- world;;
val transform :
  Device -> 'a -> Matrix ->
    (Device -> 'a -> 'b -> unit) -> 'b -> unit
```

The draw_triangles and transform functions can be used to draw an arbitrary scene:

```
> let rec draw (device : Device) color = function
    | Triangles tris ->
        Array.concat (List.map (triangle color) tris)
        |> draw_triangles device
    | Color(color, scene) -> draw device color scene
    | Transform(m, scene) ->
        transform device color m draw scene
    | Group ts -> List.iter (draw device color) ts;;
val draw : Device -> Color -> scene -> unit
```

Note that the draw function passes itself to the transform combinator in order to recurse through a Transform node in the scene graph.

In most cases, this library will be unnecessarily inefficient. However, modern graphics hardware is very fast and will be more than capable of rendering quite complicated scenes using this library.

7.2.5 Spawning visualizations from the F# interactive mode

Declarative scene graphs are most useful if the F# interactive mode is made to spawn a new window visualizing the scene whenever a computation results in a value of the type scene. This effect can be achieved by supplementing a running F# interactive session with pretty printers that spawn a separate thread to handle the visualization of a value. Threading is essential here to ensure that separate visualizations run concurrently. If visualizations are not spawned on separate threads then unwanted interactions between visualizations will occur, such as only the window in focus being refreshed. Threading is discussed in much more detail in section 9.3.

The following function spawns a window visualizing a scene:

```
> let printer scene =
    let thread =
      new System.Threading.Thread(fun () ->
        (fun device -> draw device Color.Black scene)
        |> show3d
        |> make_viewer "F# visualization"
        |> Application.Run)
    thread.SetApartmentState(ApartmentState.STA)
    thread.Start()
    "<scene>";;
val animated_printer : (unit -> scene) -> string
```

The first definition nested inside this printer function defines a variable thread that is an unstarted thread. When the thread is started (in the penultimate line), a new concurrent thread of execution will be created and the body of the anonymous function that was passed to the Thread constructor will be evaluated in that new thread of execution. This body creates a new Viewer object and uses the Run member to start a Windows Forms message loop and run the form as an application.

This formulation is *absolutely essential* for the correct working of the printer function because a Windows form must only be accessed directly from the thread on which it was created. If the form is accidentally created on the current thread but accessed on the new thread (including being passed to the Run member) then the program will not achieve the desired effect.

The next line requests that the new thread uses single-threaded apartment state, which is a requirement of a Windows form. The thread is then started and the "scene" string is returned for the interactive session to print.

The printer function can be registered with an F# interactive session using the AddPrinter method of the fsi object:

```
> fsi.AddPrinter(printer);;
```

Any expressions entered into the F# interactive mode that evaluate to a value of the type scene such that the F# interactive mode tries to print the value now results in a new window being spawned that visualizes the value.

Entering a value of the type scene now spawns a new window visualizing the scene, such as the black triangle in the following example:

```
> Triangles
    [ vec(-1.0, -1.0, 0.0),
      vec(0.0, 1.0, -1.0),
      vec(1.0, -1.0, 0.0) ];;
val it : scene = <scene>
```

Writing functions to generate and manipulate scene graphs is much easier than writing correct programmatic renderers. The remainder of this chapter demonstrates some of the many ways that this simple library can be leveraged to produce useful visualizations.

7.3 TESSELATING OBJECTS INTO TRIANGLES

In the general case, mathematically-simple objects, such as spheres, are remarkably difficult to render on modern graphics hardware.

When visualizing a biological molecule, each atom might be represented by an individual sphere. There are likely to be tens of thousands of such spheres in a single image. Consequently, each sphere must be decomposed into only a few triangles, or the graphics system will be overwhelmed and the visualization will be too slow. A simple tesselation, such as an icosahedron may well suffice in this case.

In contrast, a cartographic application might use a single sphere to represent an entire planet. The single sphere must be decomposed into many triangles for the resulting tesselation to be accurate enough to give the illusion of being a sphere. However, the sphere cannot be uniformly subdivided, or the tesselation will contain too many triangles on the far side of the sphere and too few on the near side. Thus, this application requires an adaptive tesselation, where the sphere is decomposed into triangles as a function of the view of the sphere that is required.

In this book, we shall consider only simple, uniform tesselations that suffice for the required applications. Objects such as circles and spheres can be uniformly tesselated easily. Moreover, the task of subdividing coarse meshes to obtain more accurate meshes is ideally suited to recursive functions. As we shall see, many topics in computer graphics can be solved elegantly and succinctly in F#.

7.3.1 Spheres

Spheres can be tesselated by recursively subdividing the triangular faces of an icosahedron and pushing new vertices out onto the surface of the sphere.

First, we extract the faces of an icosahedron as a list of 3-tuples of vectors:

```
> let triangles =
    [ for i, j, k in faces ->
        vertices.[i], vertices.[j], vertices.[k] ];;
val triangles : (Vector3 * Vector3 * Vector3) list
```

The following function splits an edge by averaging the end coordinates and normalizing the result to push the new vertex out onto the surface of the unit sphere:

```
> let split_edge(p, q) =
    Vector3.Normalize((p + q) * 0.5f);;
val split_edge : Vector3 * Vector3 -> Vector3
```

The following function splits a triangular face into four smaller triangles by splitting the three edges:

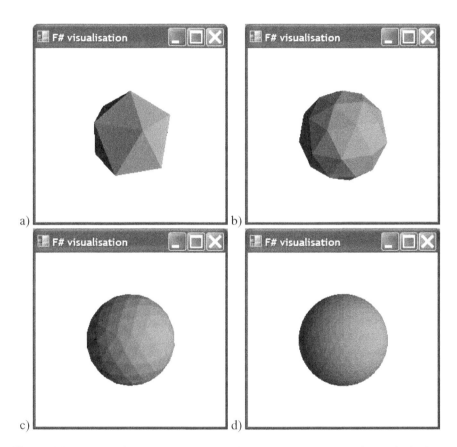

Figure 7.8   Progressively more refined uniform tesselations of a sphere, obtained by subdividing the triangular faces of an icosahedron and normalising the resulting vertex coordinate vectors to push them onto the surface of a sphere.

```
> let split_face t (a, b, c) =
    let d = split_edge(a, b)
    let e = split_edge(b, c)
    let f = split_edge(c, a)
    [a, d, f; d, b, e; e, c, f; d, e, f];;
val split_face :
  (Vector3 * Vector3 * Vector3) ->
    (Vector3 * Vector3 * Vector3) list
```

Mapping the split_face function over a sequence of triangles subdivides a coarse tesselation into a finer one:

```
> let subdivide triangles =
    List.flatten(List.map split_face triangles);;
val subdivide :
  (Vector3 * Vector3 * Vector3) list ->
    (Vector3 * Vector3 * Vector3) list
```

Finally, the following function returns the nth tesselation level of a sphere as a list of triples of vertex coordinates using the nest combinator from section 6.1.1:

```
> let sphere n =
    nest n subdivide triangles;;
val sphere : int -> (Vector3 * Vector3 * Vector3) list
```

The following expression spawns four visualizations of progressively more refined tesselations of a sphere:

```
> [ for n in 0 .. 3 ->
      Color(Color.Salmon, Triangles(sphere n)) ];;
val it : (unit -> scene) list
  = [<scene>; <scene>; <scene>; <scene>]
```

The resulting visualizations are illustrated in figure 7.8.

7.3.2   3D function plotting

A function of two variables can be tesselated into a 3D mesh of triangles using a variety of different techniques. The simplest approach is to uniformly sample the function over a grid and convert each grid square into a pair of triangles.

The following higher-order plot function samples a function f over a continuous range $[x_0, x_1]$ and $[z_0, z_1]$ to generate a scene graph ready for rendering:

```
> let plot n f x0 x1 z0 z1 =
    let g i j =
      let x = x0 + (x1 - x0) * float i / float n
      let z = z0 + (z1 - z0) * float j / float n
      vec(x, f x z, z)
    [ for i in 0 .. n-1
```

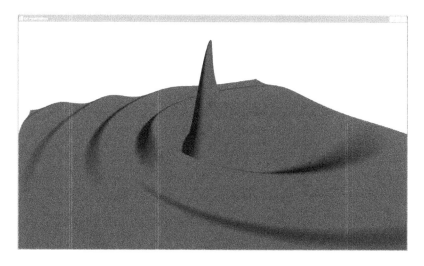

Figure 7.9   3D surface plot of $y = \sin(r + 3x)/r$ where $r = 5\sqrt{x^2 + z^2}$.

```
        for j in 0 .. n-1 ->
          let h n m = g (i + n) (j + m)
          [ h 0 0, h 0 1, h 1 1; h 0 0, h 1 1, h 1 0 ] ]
    |> List.flatten
    |> Triangles;;
val plot :
  int -> (float -> float -> float) -> float -> float ->
    float -> float -> scene
```

The following example function represents $f(x, z) = \frac{1}{r}\sin(r + 3x)$ where $r = 5\sqrt{x^2 + z^2}$:

```
> let f x z =
    let r = 5.0 * sqrt(x * x + z * z)
    sin(r + 3.0 * x) / r;;
val f : float -> float -> float
```

This function f can be tesselated and visualized by applying it to the plot function with suitable parameters:

```
> Color(Color.Red, plot 255 f -3.0 3.0 -3.0 3.0);;
val it : unit -> scene = <scene>
```

The result is illustrated in figure 7.9.

The samples provided in this chapter illustrate the basic use of Windows Forms and Managed DirectX for visualization. This allows simple visualizations to be created and even spawned independently from F# interactive sessions. However, the construction of a library capable of abstracting away the complexity involved in optimizing scene graphs and all threading issues when simulations are to be performed concurrently with visualizations is substantially more difficult. Fortunately, this has already been done by the commercial "F# for Visualization" library from Flying Frog Consultancy. One of the complete examples with visualizations from chapter 12 will use this library.

CHAPTER 8

OPTIMIZATION

Thanks to advances in computer technology, performance is no longer a concern for the majority of programs. However, many applications exist, particularly in the context of scientific computing, where performance is still important. In such cases, programs can be *optimized* to run faster by exploiting knowledge about the relative performance of different approaches.

This chapter examines the most important techniques for optimizing F# programs. The overall approach to whole program optimization is to perform each of the following steps in order:

1. Profile the program compiled with automated optimizations and running on representative input.

2. Of the sequential computations performed by the program, identify the most time-consuming one from the profile.

3. Calculate the (possibly asymptotic) algorithmic complexity of this bottleneck in terms of suitable primitive operations.

4. If possible, manually alter the program such that the algorithm used by the bottleneck has a lower asymptotic complexity and repeat from step 1.

F# for Scientists. By Jon Harrop
Copyright © 2008 John Wiley & Sons, Inc.

5. If possible, modify the bottleneck algorithm such that it accesses its data structures less randomly to increase cache coherence.

6. Perform low-level optimizations on the expressions in the bottleneck.

The mathematical concept of asympotic algorithmic complexity (covered in section 3.1) is an excellent way to choose a data structure when inputs will be large, i.e. when the asymptotic approximation is most accurate. However, many programs perform a large number of computations on small inputs. In such cases, it is not clear which data structure or algorithm will be most efficient and it is necessary to gather quantitative data about a variety of different solutions in order to justify a design decision.

In its simplest form, performance can be quantified by simply measuring the time taken to perform a representative computation.

We shall now examine different ways to measure time and profile whole programs before presenting a variety of fundamental optimizations that can be used to improve the performance of many F# programs.

8.1 TIMING

Before detailing the functions provided by F# and .NET to measure time, it is important to distinguish between two different kinds of time that can be measured.

Absolute time or *real* time, refers to the time elapsed in the real world, outside the computer.

CPU time is the amount of time that CPUs have spent performing a computation.

As a CPU's time is typically divided between many different programs, CPU time often passes more slowly than absolute time. For example, when two computations are running on a single CPU, each computation will see CPU time pass at half the rate of real time. However, if several CPUs collaborate to perform computation then the total CPU time taken may well be longer than the real time taken.

The different properties of absolute- and CPU-time make them suited to different tasks. When animating a visualization where the scene is a function of time, it is important to use absolute time otherwise the speed of the animation will be affected by other programs running on the CPU. When measuring the performance of a function or program, both absolute time and CPU time can be useful measures.

8.1.1 Absolute time

The .NET class `System.Diagnostics.Stopwatch` can be used to measure elapsed time with roughly millisecond (0.001s) accuracy. This can be productively factored into a `time` combinator that accepts a function `f` and its argument `x` and times how long `f` takes to run when it is applied to `x`, returning a 2-tuple of the time taken and the result of `f x`:

```
> let time f x =
    let timer = new System.Diagnostics.Stopwatch()
    timer.Start()
    try f x finally
    printf "Took %dms" timer.ElapsedMilliseconds;;
val time : ('a -> 'b) -> 'a -> float * 'b
```

CPU time can be measured using a similar function.

8.1.2   CPU time

The F# Sys.time function returns the CPU time consumed since the program began, and it can also be productively factored into a curried higher-order function cpu_time:

```
> let cpu_time f x =
    let t = Sys.time()
    try f x finally
    printf "Took %fs" (Sys.time() -. t);;
val cpu_time : ('a -> 'b) -> 'a -> float * 'b
```

CPU time can be measured with roughly centisecond (0.01s) accuracy using this function.

8.1.3   Looping

Many functions execute so quickly that they take an immeasurably small amount of time to run. In such cases, a higher-order loop function that repeats a computation many times can be used to provide a more accurate measurement:

```
> let rec loop n f x =
    if n > 0 then
      f x |> ignore
      loop (n - 1) f x;;
val loop : int -> ('a -> 'b) -> 'a -> 'b
```

For example, extracting the 11th element of a list takes a very short amount of time:

```
> time (List.nth 10) [1 .. 100];;
Took 0ms
val it : int = 11
> time (loop 1000000 (List.nth 10)) [1 .. 100];;
Took 567ms
val it : int = 11
```

The former timing only allows us to conclude that the real time taken to fetch the 10th element is < 0.001ms. The latter timing over a million repetitions allows us to conclude that the time taken is around $0.567\mu s$. The latter result is still erroneous because efficiency is often increased by repeating a computation (i.e. the first computation is likely to take longer than the rest) and the measurement did not account for the time spent in the `loop` function itself.

The time spent in the loop function can also be measured and turns out to be only 3% in this case:

```
> time (loop 1000000 (fun () -> ())) ();;
Took 17ms
val it : int = 11
```

The difference between real- and CPU-time is best elucidated by example.

8.1.4 Example timing

The `time` and `cpu_time` combinators can be used to measure the absolute- and CPU-time required to perform a computation. In order to highlight the slower-passing of CPU time, we shall wrap the measurement of absolute time in a measurement of CPU time. Building a set from a sequence of integers is a suitable computation:

```
> cpu_time (time Set.of_seq) (seq {1 .. 1000000});;
Took 4719ms
Took 4.671875s
val it : Set<int> = seq [1; 2; 3; 4; ...]
```

The inner timing shows that the conversion of a million-element list into a set took 4.791s of real time. The outer timing shows that the computation *and the inner timing* consumed 4.67s of CPU time. Note that the outer timing returned a shorter time span than the inner timing, as CPU time passed more slowly than real time because the CPU was shared among other programs.

Timing a variety of equivalent functions is an easy way to quantify the performance differences between them. Choosing the most efficient functions is then the simplest approach to writing high-performance programs. However, timing individual functions by hand is no substitute for the complete profiling of the time spent in all of the functions of a program.

8.2 PROFILING

Before beginning to optimize a program, it is vitally important to profile the program running on representative inputs in order to ascertain quantitative information on any bottlenecks in the flow of the program.

8.2.1 8-queens problem

Although profiling is most useful when optimizing large programs that consist of many different functions, it is instructive to look at a simple example. Consider a program to solve the 8-queens problem. This is a logic problem commonly solved on computer science courses. The task is to find all of the ways that 8 queens can be placed on a chess board such that no queen attacks any other.

As this program makes heavy use of lists, it is useful to open the namespace of the List module:

```
> open List;;
```

A function `safe` that tests whether one position is safe from a queen at another position (and vice versa) may be written:

```
> let rec safe (x1, y1) (x2, y2) =
    x1 <> x2 && y1 <> y2 &&
      x2 - x1 <> y2 - y1 && x1 - y2 <> x2 - y1;;
val safe : int * int -> int * int -> bool
```

A list of the positions on a $n \times n$ chess board may be generated using a list comprehension:

```
> let ps n =
    [ for i in 1 .. n
        for j in 1 .. n ->
          i, j ];;
val ps : int -> (int * int) list
```

A solution, represented by a list of positions of queens, may be printed to the console by printing a character for each position on a row followed by a newline, for each row:

```
> let print n qs =
    for x in 1 .. n do
      for y in 1 .. n do
        printf "%s" (if mem (x, y) qs then "Q" else ".")
      printf "\n";;
val print : int -> (int * int) list -> unit
```

Solutions may be searched for by considering each safe position `ps` and recursing twice. The first recursion searches all remaining positions, returning the accumulated solutions `accu`. The second recursion adds a queen at the current position `q` and filters out all positions attacked by the new queen before recursing:

```
> let rec search f n qs ps accu =
    match ps with
    | [] when length qs = n -> f qs accu
    | [] -> accu
    | q::ps ->
        search f n qs ps accu;;
        |> search f n (q::qs) (filter (safe q) ps);;
val search :
  ((int * int) list -> 'a -> 'a) -> int ->
```

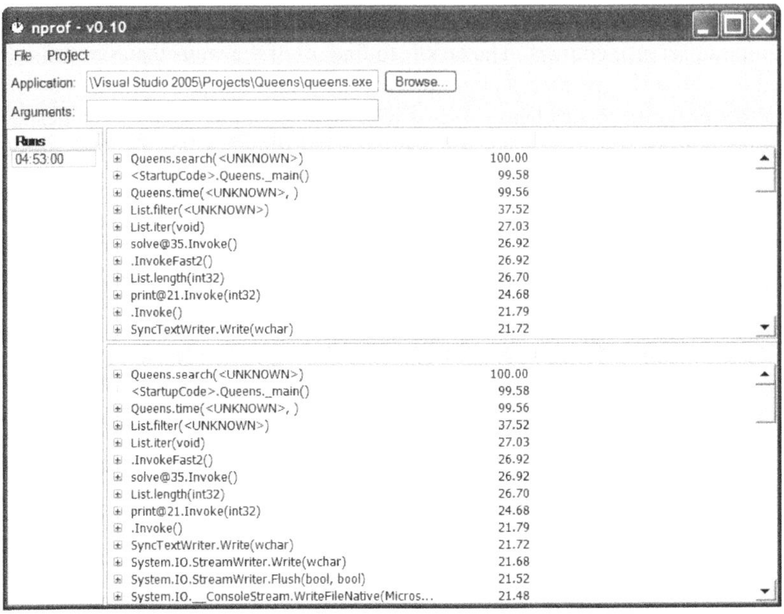

Figure 8.1 Profiling results generated by the freely-available NProf profiler for a program solving the queens problem on an 11×11 board.

```
(int * int) list -> (int * int) list -> 'a -> 'a
```

The `search` function is the core of this program. The approach used by this program, to recursively filter invalid solutions from a set of possible solutions, is commonly used in logic programming and is the foundation of some languages designed specifically for logic programming such as Prolog.

All solutions for a given board size n can be folded over using the `search` function. In this case, the fold prints each solution and increments a counter, finally returning the number of solutions found:

```
> let solve n =
    let f qs i =
      print n qs
      i + 1
    search f n [] (ps n) 0;;
val solve : int -> int
```

The problem can be solved for $n = 8$ and the number of solutions printed to the console (after the boards have been printed by the fold) with:

```
> printf "%d solutions\n" (solve 8);;
```

This program takes only 0.15s to find and print the 92 solutions for $n = 8$. To gather more accurate statistics in the profiler, it is useful to choose a larger n. Using

$n = 11$, compiling this program to an executable and running it from the NProf profiler gives the results illustrated in figure 8.1.

The profiling results indicate that almost 30% of the running time of the whole program is spent in the List.length function. The calls to this function can be completely avoided by accumulating the length nqs of the current list qs of queens as well as the list itself:

```
> let rec search f n nqs qs ps accu =
    match ps with
    | [] when nqs = n -> f qs accu
    | [] -> accu
    | q::ps ->
        search f n nqs qs ps accu
        |> search f n (nqs + 1) (q::qs)
            (filter (safe q) ps);;
val search :
  ((int * int) list -> 'a -> 'a) -> int -> int ->
    (int * int) list -> (int * int) list -> 'a -> 'a
```

The solve function just initializes with zero length nqs as well as the empty list:

```
> let solve n =
    let f qs i =
      print n qs;
      i + 1
    search f n 0 [] (ps n) 0;;
val solve : int -> int
```

This optimization reduces the time taken to compute the $2,680$ solutions for $n = 11$ from 34.8s to 23.6s, a performance improvement of around 30% as expected.

Profiling can be used to identify the performance critical portions of whole programs. These portions of the program can then be targeted for optimization. Algorithmic optimizations are the most important set of optimizations.

8.3   ALGORITHMIC OPTIMIZATIONS

As we saw in chapter 3, the choice of data structure and of algorithm can have a huge impact on the performance of a program.

In the context of program optimization, intuition is often terribly misleading. Specifically, given the profile of a program, intuition often tempts us to perform low-level optimizations on the function or functions that account for the largest proportion of the running time of the program. Counter intuitively, the most productive optimizations often stem from attempts to reduce the number of calls made to the performance-critical functions, rather than trying to optimize the functions themselves. Programmers must always strive to "see the forest for the trees" and perform algorithmic optimizations before resorting to low-level optimizations.

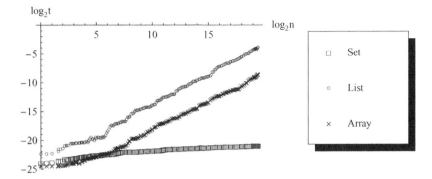

Figure 8.2   Measured performance (time t in seconds) of mem functions over set, list and array data structures containing n elements.

If profiling shows that most of the time is spent performing many calls to a single function then, before trying to optimize this function (which can only improve performance by a constant factor), consider alternative algorithms and data structures which can perform the same computation whilst executing this primitive operation less often. This can reduce the asymptotic complexity of the most time-consuming portion of the program and is likely to provide the most significant increases in performance.

For example, the asymptotic algorithmic complexity (described in section 3.1) of finding an element in a container is $O(\log n)$ if the container is a set but $O(n)$ for lists and arrays (see table 3.3). For large n, sets should be much faster than lists and arrays. Consequently, programs that repeatedly test for the membership of an element in a large container should represent that container as a set rather than as a list or an array.

An extensive review of the performance of algorithms used in scientific computing is beyond the scope of this book. The current favourite computationally-intensive algorithm used to attack scientific problems in any particular subject area is often a rapidly moving target. Thus, in order to obtain information on the state-of-the-art choice of algorithm it is necessary to refer to recently published research in the specific area.

Only once all attempts to reduce the asymptotic complexity have been exhausted should other forms of optimization be considered. We shall consider such optimizations in the next section.

8.4   LOWER-LEVEL OPTIMIZATIONS

The relative performance of most data structures is typically predicted correctly by the asymptotic complexities for $n > 10^3$. However, program performance is not always limited by a relatively small number of accesses to a large container. When performance is limited by many accesses to a small container. In such cases, $n \ll 10^3$ and the asymptotic complexity does not predict the relative performance of different kinds of container. In order to optimize such programs the programmer must use a database of benchmark results as a first indicator for design and then optimize the implementation by profiling and evolving the program design to use the constructs that turn out to be most efficient in practice.

As F# is an unusual language with unusual optimization properties, we shall endeavour to present a wealth of benchmark results illustrating practically important "swinging points" between trade-offs.

The predictability of memory accesses is an increasingly important concern when optimizing programs. This aspect of optimization is becoming increasingly important because CPU speed is increasing much more quickly than memory access speed and, consequently, the cost of stalling on memory access is growing in terms of the amount of computation that could have been performed on-CPU in the same time. In a high-level language like F#, the machine is abstracted away and the programmer is not supposed to be affected by such issues but F# is also a high-performance language and, in order to leverage this performance, it can be useful to know why different trade-offs occur in terms of the underlying structures in memory.

The remainder of this chapter provides a great deal of information, including quantitative performance measurements, that should help F# programmers to optimize performance-critical portions of code.

8.4.1 Benchmarking data structures

The point at which asymptotic complexity ceases to be an accurate predictor of relative performance can be determined experimentally. This section presents experimental results illustrating some of the more important trade-offs and quantifying relevant performance characteristics on a test machine[18].

Measuring the performance of functions in any setting, other than those in which the functions are to be used in practice, can easily produce misleading results. Although we have made every attempt to provide independent performance measurements, effects such as the requirements put upon the garbage collector by the different algorithms are always likely to introduce systematic errors. Consequently, the performance measurements which we now present must be regarded only as indicative measurements.

Figure 8.2 demonstrated that asymptotic algorithmic complexity is a powerful indicator of performance. In that case, testing for membership using `Set.mem` was found to be over $10^6\times$ faster than `List.mem` for $n = 10^6$. However, the relative performance of lists, arrays and sets was not constant over n and, in particular, arrays were the fastest container for small n. Figure 8.3 shows the ratio of the times taken to test for membership in small containers. For $n < 35$, arrays are up to twice as fast as sets.

---

[18]The test machine is a dual core 2.2GHz AMD Athlon64, 400MHz FSB and 2Gb of RAM.

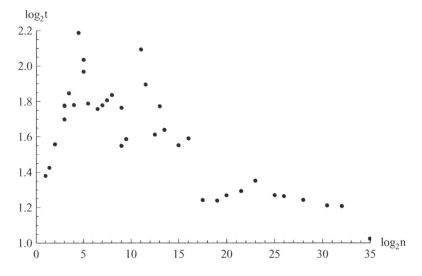

Figure 8.3   Relative time taken $t = t_S/t_a$ for testing membership in a set (t_S) and an array (t_a) as a function of the number of elements n in the container, showing that arrays are up to $2\times$ faster for $n < 35$.

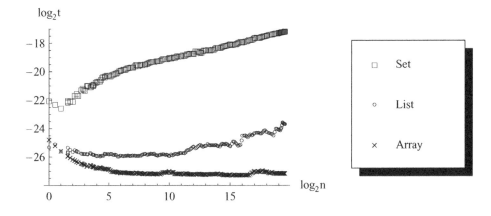

Figure 8.4   Measured performance (time t in seconds per element) of `List.of_array`, `Array.copy` and `Set.of_array` data structures containing n elements.

As this example illustrates, benchmarking is required when dealing with many operations on small containers.

Figure 8.4 illustrates the time taken to create a list, array or set from an array. Array creation is fastest, followed by list creation and then set creation. For 10^3 elements on the test machine, creating a list element takes 16ns, an array element takes 6ns and a set element takes 1.9μs. Moreover, array creation takes an approximately-constant

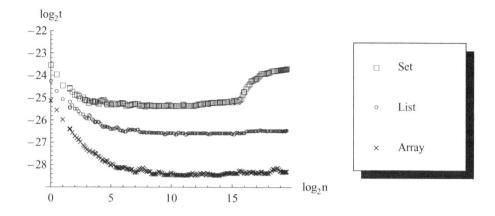

Figure 8.5   Measured performance (time t in seconds per element) of `iter` functions over list, array and set data structures containing n elements.

time per element regardless of the size of the resulting array whereas list and set creation takes longer per element for larger data structures, i.e. the total time taken to create an n-element list or set is not linear in n.

Although tests for membership and the set theoretic operations (union, intersection and difference) are asymptotically faster for sets than for lists and arrays, the fact that set creation is likely to be at least $300\times$ slower limits their utility to applications that perform many elements lookups for every set creation. In fact, we shall study one such application of sets in detail in chapter 12.

Figure 8.5 illustrates the time taken to iterate over the elements of lists, arrays and sets. Iterating over arrays is fastest, followed by lists and then sets.

Iterating over a list is $4\times$ slower than iterating over an array. The performance degradation for lists and sets when n is large is a cache coherency issue. Like most immutable data structures, both lists and sets allocate many small pieces of memory that refer to each other. These tend to scatter across memory and, consequently, memory access becomes the bottleneck when the list or set does not fit on the CPU cache. In contrast, arrays occupy a single contiguous portion of memory, so sequential access (e.g. `iter`) is faster for arrays. For example, iterating over a set is $8\times$ slower than iterating over an array for $n = 10^3$ and $15\times$ slower for $n = 10^6$.

Figure 8.6 illustrates the time taken to perform a left fold, summing the elements of a list, array or set in forward order. The curves are qualitatively similar to figure 8.5 for the `iter` functions but `fold_left` is up to $10\times$ slower than `iter`.

Figure 8.7 illustrates the time taken to perform a right fold, summing the elements of a list, array or set. Left and right folds perform similarly for arrays and sets but lists show slightly different behaviour due to the stacking of intermediate results required to traverse the list in reverse order. Specifically, `fold_right` is 30% slower than `fold_left` for lists.

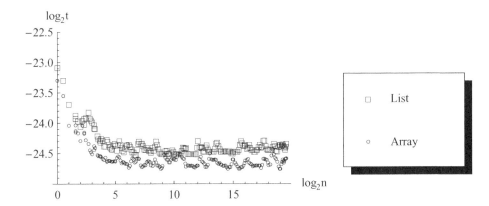

Figure 8.6 Measured performance (time t in seconds per element) of the `fold_left` functions over list and array data structures containing n elements.

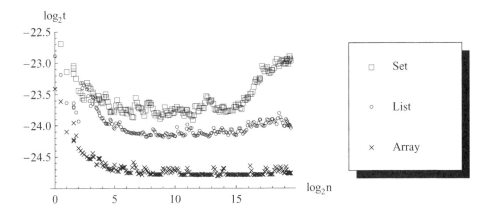

Figure 8.7 Measured performance (time t in seconds per element) of the `fold_right` functions over list, array and set data structures containing n elements.

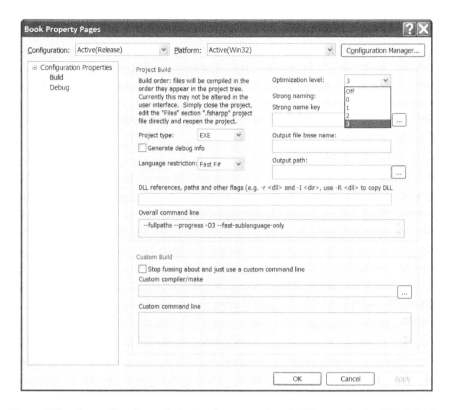

Figure 8.8   Controlling the optimization flags passed to the F# compiler by Visual Studio 2005.

These benchmark results may be used as reasonably objective, quantitative evidence to justify a choice of data structure. We shall now examine other forms of optimization, in approximately-decreasing order of productivity.

8.4.2 Compiler flags

The simplest way to improve the performance of an F# program is to instruct the F# compiler to put more effort into optimizing the code. This is done by providing an optimization flag of the form -Ooff, -O1, -O2 or -O3 where the latter instructs the compiler to try as hard as possible to improve the performance of the program.

In Visual Studio, the optimization setting for the compiler is controlled from the Project ▷ *project* Properties... dialog box (illustrated in figure 8.8).

As a last resort, program transformations performed manually should be considered as a means of optimization. We shall now examine several different approaches. Although we try to associate quantitative performance benefits with the various approaches, these are only indicative and are often chosen to represent the best-case.

8.4.3 Tail-recursion

Straightforward recursion is very efficient when used in moderation. However, the performance of deeply recursive functions can suffer. Performance degradation due to deep recursion can be avoided by performing *tail recursion* [16].

If a recursive function call is not tail recursive, state will be stored such that it may be restored after the recursive call has completed. This storing, and the subsequent retrieving, of state is responsible for the performance degradation. Moreover, the intermediate state is stored on a limited resource known as the stack which can be exhausted, resulting in StackOverflowException being raised.

Tail recursion involves writing recursive calls in a form which does not need this state. Most simply, a tail call returns the result of the recursive call directly, i.e. without performing any computation on the result.

For example, a function to sum a list of arbitrary-precision integers may be written:

```
> let rec sum = function
    | [] -> 0I
    | h::t -> h + sum t;;
val sum : bigint list -> bigint
```

Summing 10^4 elements takes $2ms$:

```
> time sum [1I .. 10000I];;
Took 2ms
val it : bigint = 50005000I
```

Attempting to sum 10^5 elements exhausts stack space:

```
> time sum [1I .. 100000I]
Process is terminated due to StackOverflowException
```

The result of the recursive call sum t in the body of the sum function is not returned directly but, rather, has h added to it to create a new value that is then returned. Thus, this recursive call to sum is not a tail call.

The sum function can be written in tail recursive form by accumulating the sum in an argument and returning the accumulator when the end of the list is reached:

```
> let rec sum_tr_aux (accu : bigint) = function
    | [] -> accu
    | h::t -> sum_tr_aux (h + accu) t;;
val sum_tr_aux : bigint -> bigint list -> bigint
```

The signature of the auxiliary function sum_tr_aux is not the same as that of sum, as it must now be initialized with an accumulator of zero. A tail recursive sum_tr function can be written in terms of sum_tr_aux by applying zero as the initial accumulator:

```
> let sum_tr list =
    sum_tr_aux 0I list;;
val sum_tr : int list -> int
```

As this `sum_tr` function is tail recursive, it will not suffer from stack overflows when given long lists:

```
> time sum_tr [1I .. 10000I]
Took 8ms
val it : bigint = 50005000I
> time sum_tr [1I .. 100000I]
Took 47ms
val it : bigint = 5000050000I
> time sum_tr [1I .. 1000000I];;
Took 196ms
val it : bigint = 500000500000I
```

As we have seen, non-tail-recursive functions are not robust when applied to large inputs. However, many functions are naturally written in a non-tail-recursive form.

For example, the built-in `List.fold_right` function is most simply written as:

```
> let rec fold_right f list accu =
    match list with
    | [] -> accu
    | h::t -> f h (fold_right f t accu);;
val fold_right : ('a -> 'b -> 'b) -> 'a list -> 'b -> 'b
```

This form is not tail recursive. The built-in `List.fold_right` function is actually tail recursive, and tries to be efficient, but it still has an overhead compared to the `List.fold_left` function (see figures 8.6 and 8.7). The reason for this overhead stems from the fact that the `fold_left` function considers the elements in the list starting at the front of the list whereas the `fold_right` function starts at the back. On a container that allows random access, such as an array, there is little difference between considering elements in forward or reverse order. However, the primitive operation used to decompose a list presents the first element and the remaining list. Consequently, looping over lists is more efficient in forward order than in reverse order.

Many standard library functions suffer from the inefficiency that results from having been transformed into the more robust tail recursive form. In the `List` module, the `fold_right`, `map`, `append` and `concat` functions are not naturally tail recursive.

8.4.4 Avoiding allocation

Although the .NET platform is based upon a common language run-time (**CLR**) that allows seamless interoperability between programs written in different languages, the platform has been optimized primarily for the C# language. As a conventional imperative language, C# programs have a typical value lifetime distribution that is significantly different to that of most functional programs. Specifically, functional programs often allocate huge numbers of short-lived small values and standalone

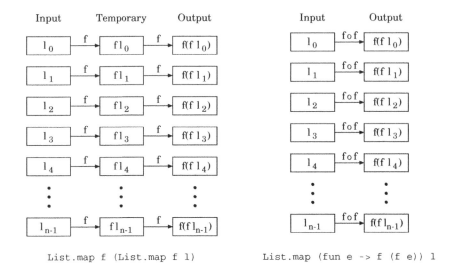

Input   Temporary   Output   Input   Output

List.map f (List.map f l)   List.map (fun e -> f (f e)) l

Figure 8.9   Deforestation refers to methods used to reduce the size of temporary data, such as the use of composite functions to avoid the creation of temporary data structures illustrated here: a) mapping a function f over a list l twice, and b) mapping the composite function $f \circ f$ over the list l once.

functional programming languages like OCaml are optimized for this [11]. As .NET is not optimized for this, allocation in F# is more expensive and, consequently, optimizations that avoid allocation are correspondingly more valuable when trying to make F# programs run more quickly.

8.4.4.1 *Deforesting*   Functional programming style often results in the creation of temporary data due to the repeated use of maps, folds and other similar functions. The reduction of such temporary data is known as *deforestation*. In particular, the optimization of performing functions sequentially on elements rather than containers (such as lists and arrays) in order to minimize the number of temporary containers created (illustrated in figure 8.9).

For example, the Shannon entropy H of a vector **v** representing a discrete probability distribution is given by:

$$H(\mathbf{v}) = \sum_{i=1}^{n} v_i \ln |v_i|$$

This could be written in F# by creating temporary containers, firstly $u_i = \ln v_i$ and then $w_i = u_i v_i$ and finally calculating the sum $H(\mathbf{v}) = \sum_i w_i$:

```
> open List;;
> let entropy1 v =
    fold_left ( + ) 0.0 (map2 ( * ) v (map log v));;
```

```
val entropy1 : float list -> float
```

This function is written in an elegant, functional style by composing functions.

However, the subexpressions map log v for u and map2 (*) v (...) for w both result in temporary lists. Creating these intermediate results is slow. Specifically, the $O(n)$ allocations of all of the list elements in the intermediate results is much slower than any of the arithmetic operations performed by the rest of this function.

Fortunately, this function can be completely deforested by performing all of the arithmetic operations at once for each element, combining the calls to +, * and log into a single anonymous function:

```
> let entropy2 v =
    fold_left (fun h v -> h + v * log v) 0.0 v;;
val entropy2 : float list -> float
```

Note that this function produces no intermediate lists, i.e. it performs only $O(1)$ allocations.

The deforested entropy2 function is much faster than the entropy1 function. For example, it is $\sim 14\times$ faster when applied to a 10^7-element list:

```
> time entropy1 [1.0 .. 10000000.0];;
Took 13971ms
val it : float = 7.809048631e+14
> time entropy2 [1.0 .. 10000000.0];;
Took 1045ms
val it : float = 7.809048631e+14
```

Deforesting can be a very productive optimization when using eager data structures such a lists, arrays, sets, maps, hash tables and so on. However, the map2 and map functions provided by the Seq module are lazy: rather than building an intermediate data structure they will store the function to be mapped and the original data structure. Thus, a third variation on the entropy function is deforested and lazy:

```
> let entropy3 v =
    Seq.map2 ( * ) v (Seq.map log v)
    |> Seq.fold ( + ) 0.0;;
val entropy3 : #seq<float> -> float
```

The performance of a lazy Seq-based implementation is between that of the deforested and forested eager implementations:

```
> time entropy3 [1.0 .. 10000000.0];;
Took 6140ms
val it : float = 7.809048631e+14
```

Lazy functions are usually slower than eager functions so it may be surprising that the lazy entropy3 function is actually faster than the eager entropy1 function in this case. The reason is that allocating large data structures is proportionally much slower than allocating small data structures. As the `Seq.map` function is lazy, the `entropy3` function never allocates a whole intermediate list but, rather, allocates intermediate lazy lists. Consequently, the lazy `entropy3` function evades the allocation and subsequent garbage collection of a large intermediate data structure and is twice as fast as a consequence.

The lazy implementation can also be deforested:

```
> let entropy4 v =
    Seq.fold (fun h v -> h + v * log v) 0.0 v;;
val entropy4 : #seq<float> -> float
```

This deforested lazy implementation is faster than the forested lazy implementation but not as fast as the deforested eager implementation:

```
> time entropy4 [1.0 .. 10000000.0];;
Took 2167ms
val it : float = 7.809048631e+14
```

The deforested eager `entropy2` function is the fastest implementation because it avoids the overheads of both intermediate data structures and laziness. Consequently, this style should be preferred in performance-critical code and, fortunately, it is easy to write in F# thanks to first-class functions and the built-in function composition operators.

8.4.4.2 *Avoid copying*

Explicit declarations of compound data structures almost certainly entail allocation. Consequently, a simple approach to avoiding copying is to write performance-intensive rewrite operations so that they reuse previous values rather than rebuilding (copying) them.

For example, the following function contracts a pair where either element is zero but incurs a copy by explicitly restating the pair `x, y` even when the input is being returned as the output:

```
> let f = function
    | 0, _ | _, 0 -> 0, 0
    | x, y -> x, y;;
val f : int * int -> int * int
```

This function is easily optimized by returning the input as the output explicitly rather than using the expression `x, y`:

```
> let f = function
    | 0, _ | _, 0 -> 0, 0
    | a -> a;;
val f : int * int -> int * int
```

The second match case no longer incurs an allocation by unnecessarily copying the input pair.

Referential equality (described in section 1.4.3) can be used to avoid unnecessary copying. This typically involves checking if the result of a recursive call is referentially equal to its input.

Consider the following implementation of the insertion sort:

```
> let rec sort1 = function
    | [] -> []
    | x::xs ->
        match sort1 xs with
        | y::ys when x > y -> y::sort1(x::ys)
        | ys -> x::ys;;
val sort1 : 'a list -> 'a list
```

In general, this is an adequate sort function only for short lists because it has $O(n^2)$ complexity where n is the length of the input list. However, if the sort1 function is often applied to already-sorted lists then this function is unnecessarily inefficient because it always copies its input.

The following optimized implementation uses referential equality to spot when the list x::xs that is being reconstructed and returned will be identical to the given list and returns the input list directly when possible:

```
> let rec sort2 = function
    | [] -> []
    | x::xs as list ->
        match sort2 xs with
        | y::ys when x > y -> y::sort2(x::ys)
        | ys -> if xs == ys then list else x::ys;;
val sort2 : 'a list -> 'a list
```

This approach used by the sort2 function avoids unnecessary copying and can improve performance when the input list is already sorted or when insertions are required near the front of the list.

8.4.5   Terminating early

Algorithms may execute more quickly if they can avoid unnecessary computation by terminating as early as possible. However, the trade-off between any extra tests required and the savings of exiting early can be difficult to predict. The only general solution is to try premature termination when performance is likely to be enhanced and revert to the simpler form if the savings are not found to be significant. We shall now consider a simple example of premature termination as found in the core library as well as a more sophisticated example requiring the use of exceptions.

The for_all function in the List module is an enlightening example of an early-terminating function. This function applies a predicate function p to elements in a list, returning true if the predicate was found to be true for all elements and false otherwise. Note that the predicate need not be applied to all elements in the list, as the result is known to be false as soon as the predicate returns false for any element. The for_all function may be written:

```
> let rec for_all1 p = function
    | [] -> true
```

```
    | h::t -> p h && for_all1 p t;;
val for_all1 : ('a -> bool) -> 'a list -> bool
```

The premature termination of this function is not immediately obvious. In fact, the `&&` operator has the unusual semantics of in-order, short-circuit evaluation. This means that the expression p h will be evaluated first and *only if the result is* `true` will the expression `for_all1 p t` be evaluated. Consequently, this implementation of the `for_all` function can return `false` without recursively applying the predicate function p to all of the elements in the given list.

Moreover, the short-circuit evaluation semantics of the `&&` operator makes the above function equivalent to:

```
> let rec for_all1 p = function
    | [] -> true
    | h::t -> if p h then for_all1 p t else false;;
val for_all1 : ('a -> bool) -> 'a list -> bool
```

Consequently, these `for_all` functions are actually tail recursive (the result of the call `for_all p t` is returned without being acted upon).

These implementations of the `for_all` function are very efficient. Terminating on the first element takes under 50ns:

```
> [1 .. 1000]
  |> time (loop 100000 (for_all1 ((<>) 1)));;
Took 5ms
val it : bool = false
```

Terminating on the 1000th element takes $11\mu s$:

```
> [1 .. 1000]
  |> time (loop 100000 (for_all1 ((<>) 1000)));;
Took 1144ms
val it : bool = false
```

These performance results will now be used to compare an alternative exit strategy: exceptions.

A similar function may be written in terms of a fold:

```
> let for_all2 p list =
    List.fold_left ( && ) true list;;
val for_all : ('a -> bool) -> 'a list -> bool
```

However, this `for_all2` function will apply the predicate p to every element of the list, i.e. it will not terminate early.

The fold-based implementation can be made to terminate early by raising and catching an exception:

```
> let for_all3 p list =
    try
      let f b h = p h && raise Exit
```

```
      List.fold_left f true list
    with
    | Exit ->
        false;;
val for_all3 : ('a -> bool) -> 'a list -> bool
```

Indeed, there is now no reason to accumulate a boolean that is always `true`. The function can be written using `iter` or a comprehension instead of a fold:

```
> let for_all4 p list =
    try
      for h in list do
        if not (p h) then raise Exit
      true
    with
    | Exit ->
        false;;
val for_all4 : ('a -> bool) -> 'a list -> bool
```

This function has recovered the asymptotic efficiency of the original `for_all1` function but the actual performance of this `for_all4` function cannot be determined without measurement. Indeed, as exceptions are comparatively slow in .NET the exception-based approach is likely to be significantly slower when the exception is raised and caught.

When terminating on the first element, the performance is dominated by the cost of raising and catching an exception. The time taken to return at the first element is $29\mu s$:

```
> [1 .. 1000];;
  |> time (loop 100000 (for_all4 ((<>) 1)));;
Took 2862ms
val it : bool = false
```

Terminating on the 1000th element takes $43\mu s$:

```
> [1 .. 1000];;
  |> time (loop 100000 (for_all4 ((<>) 1000)));;
Took 4269ms
val it : bool = false
```

In the latter case, a significant amount of other computation is performed and the performance cost of the exception is not too significant. Consequently, the `for_all4` function is $4\times$ slower than the original `for_all1` function. However, if the flow involves a small amount of computation and an exception, as it does in the former case, then the exceptional route is $600\times$ slower. This is such a significant performance difference that the standard library provides alternative functions for performance critical code that avoid exceptions in all cases. For example, the `assoc` and `find` functions in the `List` module have `try_assoc` and `tryfind` alternatives that return an option result to avoid raising an exception when a result is not found.

8.4.6 Avoiding higher-order functions

In F#, applying functions as arguments can lead to slower code. Consequently, avoiding higher-order functions in the primitive operations of numerically-intensive algorithms can significantly improve performance.

For example, a function to sum an array of floating point numbers may be written in terms of a fold:

```
> let sum1 a =
    Array.fold_left ( + ) 0.0 a;;
val sum1 : float array -> float
```

Alternatively, the function may be written using an explicit loop rather than a fold:

```
> let sum2 a =
    let r = ref 0.0
    for i = 0 to Array.length a - 1 do
      r := !r + a.[i]
    !r;;
val sum2 : float array -> float
```

Clearly, the fold has significantly reduced the amount of code required to provide the required functionality.

The overhead of using a higher-order function results in the sum1 function executing significantly more slowly than the sum2 function:

```
> let a = Array.init 10000000 float;;
val a : float array

> time sum1 a;;
Took 376ms
val it : float = 4.9999995e13

> time sum2 a;;
Took 104ms
val it : float = 4.9999995e13
```

In this case, the sum2 function is $3.6\times$ faster than the sum1 function. In fact, performance can be improved even more by using a mutable value.

8.4.7 Use mutable

The mutable keyword can be used in a let binding to create a locally mutable value. Mutable values are not allowed to escape their scope (e.g. they cannot be used inside a nested closure). However, mutable values can be more efficient than references.

For example, the previous section detailed a function for summing the elements of a `float array`. The fastest implementation used a reference for the accumulator but this can be written more efficiently using a `mutable`:

```
> let sum3 a =
    let mutable r = 0.0
    for i = 0 to Array.length a - 1 do
      r <- r + a.[i]
    r;;
val sum3 : float array -> float
```

Using a mutable makes this `sum3` function $2.7\times$ faster than the fastest previous implementation `sum2` that used a reference:

```
> time sum3 a;;
Took 39ms
val it : float = 4.9999995e13
```

The combination of avoiding higher-order functions and using `mutable` makes this `sum3` function $9.6\times$ faster than the original `sum1`.

8.4.8   Specialized functions

Particularly in the context of scientific computing, F# programs can benefit from the use of specialized high-performance functions. This includes some complicated numerical algorithms for matrix and Fourier analysis, which will be covered in chapter 9, but the F# standard library also provides some fast functions for common mathematical values.

The built-in `vector` type implements arbitrary-dimensionality vectors with `float` elements and provides operators that are considerably faster than equivalent functions from the `Array` module:

```
> let a, b = [|1.0 .. 1000.0|], [|1.0 .. 1000.0|];;
val a : float array
val b : float array
> time (loop 100000 (Array.map2 ( + ) a)) b
  |> ignore;;
Took 5744ms
val it : unit = ()
> (vector a, vector b)
  |> time (loop 100000 (fun (a, b) -> a + b))
  |> ignore;;
Took 2014ms
val it : unit = ()
```

In this case, vector addition is $2.9\times$ faster using the specialized vector addition.

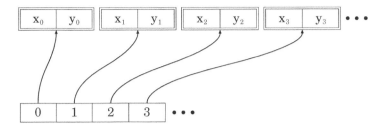

Figure 8.10   An array of tuples or records containing pairs of float values incurs a level of indirection.

| x_0 | y_0 | x_1 | y_1 | x_2 | y_2 | x_3 | y_3 | • • • |

Figure 8.11   A struct can be used to completely unbox the float values, placing them directly into the array. This is often the most efficient representation.

8.4.9   Unboxing data structures

Typically in functional languages, most values are *boxed*. This means that a value is stored as a reference to a different piece of memory. Although elegant in providing efficient referential transparency, unnecessary boxing can incur significant performance costs.

For example, much of the efficiency of arrays stems from their elements occupying a contiguous portion of memory and, therefore, accesses to elements with similar indices are cache coherent. However, if the array elements are boxed, only the references to the data structures will be in a contiguous portion of memory (see figure 8.10). The data structures themselves may be at completely random locations, particularly if the array was not filled sequentially. Consequently, cache coherency may be very poor.

For example, computing the product of an array of complex numbers is unnecessarily inefficient when the numbers are represented by a (float * float) array (see figure **??**). A struct is often more efficient as this avoids all unboxing (see figure 8.11).

A function to compute the product of an array of complex numbers might be written over the type (float * float) array, where each element is a 2-tuple representing the real and imaginary parts of a complex number:

```
> let product1 zs =
    let mutable re = 1.0
    let mutable im = 0.0
    for i = 0 to Array.length zs - 1 do
      let zr, zi = zs.[i]
      let r = re * zr - im * zi
      and i = re * zi + im * zr
      re <- r
```

```
      im <- i
   re, im;;
val product1 : (float * float) array -> float * float
```

To avoid boxing completely, the complex number can be stored in a struct:

```
> type Complex = struct
     val re : float
     val im : float
     new(x, y) = { re = x; im = y }
  end;;
```

A more efficient alternative may act upon values of the type `Complex array`:

```
> let product2 (zs : Complex array) =
     let mutable re = 1.0
     let mutable im = 0.0
     for i = 0 to Array.length zs - 1 do
       let r = re * zs.[i].r - im * zs.[i].i
       and i = re * zs.[i].i + im * zs.[i].r
       re <- r
       im <- i
     re, im;;
val product2 : Complex array -> float * float
```

Benchmarking these two functions compiled to native code with full compiler optimizations and acting upon randomly shuffled 2^{24}-element arrays, we find that the tuple-based `product1` function takes $1.94s$ to complete and the struct-based `product2` function takes only $0.25s$.

8.4.10   Eliminate needless closures

The definition of a local function or the partial application of a curried function can create a closure. This functionality is not provided by the .NET platform and, consequently, F# must distinguish between the types of ordinary functions and closures to remain compatible. The type of a closure is bracketed, so `'a -> 'b` becomes `('a -> 'b)`.

For example, both of the following functions compute the next integer and may be used equivalently in F# programs but the former is an ordinary function (like a C# function) whereas the latter is a closure that results from partial application of the curried + function:

```
> let succ1 n =
     n + 1;;
val succ1 : int -> int
> let succ2 =
     ( + ) 1;;
val succ2 : (int -> int)
```

Closures encapsulate both a function and its environment (such as any partially applied arguments). Consequently, closures are heavier than functions and the creation and invocation of a closure is correspondingly slower. In fact, the F# compiler translates a closure into an object that encapsulates the environment and function.

8.4.11 Inlining

The F# programming language provides an inline keyword that provides a simple way to inline the body of a function at its calls sites:

```
> let f x y = x + y;;
val f : int -> int -> int
> let inline f x y = x + y;;
val f : ^a -> ^b -> ^c when ^a :
  (static member (+) : ^a -> ^b -> ^c)
```

The body of the function marked inline will be inserted verbatim whenever the function is called. Note that the type of the inlined function is very different. This is because ordinary functions require ad-hoc polymorphic functions such as the + operator to be ossified to a particular type, in this case the default type int. In contrast, the inlining of the function in the latter case allows the body of the function to adopt whatever type is inferred from any context it is inserted into individually. Hence its type requires only that a + operator is defined in the context of the call site.

The performance implications of inlining vary wildly, with numeric code often benefitting from judicious inlining but the performance of symbolic code is often deteriorated by inlining.

8.4.12 Serializing

The F# input_value and output_value functions or the standard .NET *serialization* routines themselves can be used to store and retrieve arbitrary data using a binary format. However, at the time of writing these routines in the .NET library are not optimized and require as much space as a textual format and often take several times as long as a custom parser to load. In particular, the current .NET serialization implementation has quadratic complexity when serializing data structures containing many shared functions. This includes the serialization of data structures that contain INumeric implementations, such as the F# vector and matrix types.

Consequently, programs with performance limited by IO may be written more efficiently by handling custom formats, such as textual representations of data (parsed using the techniques described in chapter 5) or even using a custom binary format. CPU performance has increased much more quickly than storage performance and, as a consequence, data compression can sometimes be used to improve IO performance by reducing the amount of data being stored in exchange for a higher CPU load.

Having examined the many ways F# programs may be optimized, we shall now review existing libraries which may be of use to scientists.

CHAPTER 9

LIBRARIES

A great deal of time and effort has been put into existing software. Many libraries have resulted from years of development and the ability to reuse this work is essential. There are two different forms of library of interest to the F# programmer:

- .NET libraries
- Native libraries

The .NET platform is famous for providing a wealth of library functions, particularly in the contexts of web and GUI programming. These libraries will be of interest to F# programmers primarily for the useful functionality that they provide, such as graphing tools. Many third party libraries are available for the .NET platform and we shall endeavour to cite those that we have found to be most useful and of the highest quality.

The term *native library* refers to platform-specific libraries that typically predate the .NET platform. The few remaining native libraries that have not been succeeded by .NET libraries are desirable primarily because they offer exceptionally high performance.

This chapter describes built-in functionality provided with the .NET platform as well as the use of both .NET and native libraries.

F# for Scientists. By Jon Harrop
Copyright © 2008 John Wiley & Sons, Inc.

9.1   LOADING .NET LIBRARIES

On the .NET platform, extra functionality is typically provided in the form of a *Dynamically Linked Library* (DLL). These files have the suffix ".dll" and can be loaded into an F# interactive session or referenced from a compiled F# program using the #r directive. For example, the following loads the "XYGraph.dll" library:

```
> #r "XYGraph.dll";;
```

A set of directories is searched for this file and F# complains if it fails to find the DLL. The search can be expanded to include other directories using the #I directive. For example, the following adds the directory "C:\Program Files" to the search path:

```
> #I @"C:\Program Files";;
```

Compiled programs can also specify search paths and DLLs using -I and -r compile-line arguments.

9.2   CHARTING AND GRAPHING

Although the .NET platform does not bundle any charting or graphing tools there are a wide variety of libraries available, both open source and commercial products.

We have found the ComponentXtra libraries[19] to be simple to use and powerful enough to be useful in many situations. Moreover, their tools are free except for the saving and printing capabilities. The XYGraph class draws a 2D graph into a Windows form.

An object of the class XYGraph is a Windows Forms control and can be added into a form:

```
> let form = new Form(Text="Sine wave", Visible=true);;
val form : Form

> let graph =
    new componentXtra.XYGraph(Dock=DockStyle.Fill);;
val graph : componentXtra.XYGraph

> form.Controls.Add(graph);;
```

The labels on the graph are properties that can be set:

```
> graph.XtraTitle <- "Sine wave";;
> graph.XtraLabelX <- "x";;
> graph.XtraLabelY <- "sin x";;
```

A series can be added to the graph using the AddGraph method:

```
> let g =
```

---

[19]http://www.componentxtra.com

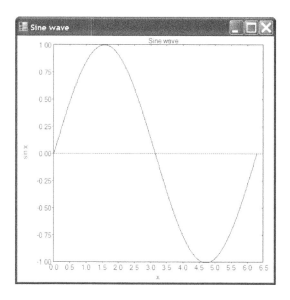

Figure 9.1   The XYGraph tool from ComponentXtra.

```
    let dash = Drawing2D.DashStyle.Solid
    graph.AddGraph("", dash, Color.Red, 1, false);;
val g : int
```

The int result is a handle to this series. Points can be added to the series by passing this handle and the new coordinates to the AddValue method:

```
> for x in 0.0f .. 0.01f .. 6.29f do
    graph.AddValue(g, x, sin y);;
```

Finally, the graph must be redrawn with the new points:

```
> graph.DrawAll();;
```

The resulting 2D graph is illustrated in figure 9.1.

9.3 THREADS

Many applications can benefit from the ability to execute threads of program concurrently. On the .NET platform, this functionality is provided by *threading*.

The design of concurrent programs and the use of threading on the .NET platform is a huge topic fraught with perils. Consequently, this chapter aims only to provide the reader with a basic understanding of threading on the .NET platform and to provide some useful higher-order functions that can be used to improve the performance of F# programs whilst avoiding as many pitfalls as possible.

Concurrency serves three different uses that have different properties and, consequently, require different support:

1. CPU bound computation, e.g. dividing a long running computation between two CPUs to double performance.

2. Non-CPU bound computation, e.g. downloading many protein sequences over the internet without wasting time waiting for one download to finish before beginning another.

3. Interactivity, e.g. using a multithreaded GUI to ensure that the user interface of an application continues to function while the application is processing data.

CPU-bound computations are best suited to one thread for each CPU whereas non-CPU bound computations benefit from having a number of threads that is unrelated to the number of CPUs, chosen to be a trade-off between sufficient parallelism and the overhead of context switching between threads. Multithreaded GUIs require more sophisticated use of low-level threading constructs.

The following sections begin by detailing the basic use of .NET threads before presenting functions to handle both CPU-bound and non-CPU-bound computations safely and easily.

9.3.1   Thread safety

Determinism is a vitally important property of most programs. In the context of threading, a function that is careful to ensure determinism in the context of parallel execution is referred to as being *thread safe*. Designing functions to be thread safe is arbitrarily difficult and, consequently, containing this complexity is essential if large programs are to run correctly.

Consider a program that spawns two threads that execute in parallel, adjusting a counter. The first thread decrements the counter and the second thread increments the counter. All things being equal, we might expect the counter to be either -1, 0 or 1 at any given time. However, this is not the case for two reasons:

- CPU time may not be divided equally between the two threads by the OS.

- Increment and decrement are not *atomic* operations, so the value of the counter may be adjusted by one thread while the other thread is in the middle of an increment or decrement.

The unequal division of CPU time can be accounted for by queuing operations for execution rather than dividing them equally between threads. Each thread of computation executes tasks from the queue. If one thread is given more CPU time then it will consume more tasks from the queue, keeping all CPUs busy and making more efficient use of CPU time.

The latter problem is more subtle and can be solved using some low-level constructs. Specifically, threads can be synchronized to ensure that certain operations (increment and decrement in this case) are executed atomically by only one thread at a time. This is achieved by wrapping the non-atomic operation in a *lock*, a low-level construct that excludes other threads from executing the code at the same time. In this case, the increment and decrement would be wrapped in locks that depended upon a single value, thus precluding one thread from incrementing while the other was decrementing and vice-versa.

The remainder of this section introduces the fundamental threading constructs provided by the .NET platform before describing how useful higher-order functions can be composed from these constructs and reused in F# programs to simplify the use of concurrency in many scientific programs.

9.3.2 Basic use

Threading is provided via classes from the `System.Threading` namespace:

```
> open System.Threading;;
```

Concurrent threads of computation can be spawned by constructing a `Thread` object with a function f to execute and calling its `Start` method to begin execution. The following higher-order `spawn` function begins executing the given function f concurrently on a separate thread:

```
> let spawn (f : unit -> unit) =
    let thread = new Thread(f)
    thread.Start()
    thread;;
val spawn : (unit -> unit) -> Thread
```

After the `spawn` function is invoked, another computation is running concurrently. In most cases, the main thread will want to pause until the spawned thread completes, e.g. to read the result of the completed thread. This is achieved by calling the `Join` method of the `Thread` object that was returned by the `spawn` function.

The following `execute` function spawns several threads and waits for them all to complete before returning:

```
> let execute n f =
    [| for i in 1 .. n ->
        spawn f |]
    |> Array.iter (fun t -> t.Join());;
val execute : int -> (unit -> unit) -> unit
```

In the context of functional programming, the single most important improvement that can be made to the `spawn` and `execute` functions is to create variants that allow values fed through functions that execute concurrently. This requires the type of the function f to be generalized from `unit -> unit` to `'a -> 'b`.

This improvement hints at what is perhaps the single most useful parallel programming construct in scientific computing: the parallel higher-order map function. This function applies a given function to each element of an array, executing applications concurrently.

A naive implementation of a parallel map can be written using the minimal threading functionality already discussed:

```
> let map f a =
    let b = Array.map (fun _ -> None) a
    let f i x = spawn (fun () -> b.[i] <- Some(f x))
    for thread in Array.mapi f a do
      thread.Join()
    Array.map Option.get b;;
val map : ('a -> 'b) -> 'a array -> 'b array
```

This implementation spawns one thread for each element of the input array a. Each thread fills the corresponding element of the array b with its result. As the type of the result is not known, the intermediate array b contains option types that are initially set to None. All of the threads are started and then all of the threads are joined, pausing the main thread until every spawned thread has completed. When all of the threads have completed the array b contains only Some values, which are extracted using the Option.get function to return the final result.

There is a trade-off between the performance improvement due to concurrent execution distributed across more than one CPU and the overheads involved in this parallel map function:

- Boxing and unboxing of the option type.

- Thread creation and handling.

- Context switching if there are more threads than free CPUs.

- Repeated iteration over the input array.

However, even this naive implementation can give a significant performance improvement when the input array contains one element for each CPU, the computations take roughly-equal time and the time taken is much longer than the overheads.

For example, computing the 40th Fibonacci number using the fib function (given at the beginning of section 6.1.4) takes 2.6s:

```
> time fib 40;;
Took 2606ms
val it : int = 102334155
```

Performing the same computation twice using the sequential Array.map function takes just over twice as long, as expected:

```
> time (Array.map fib) [|40; 40|];;
Took 5336ms
val it : int array = [|102334155; 102334155|]
```

Exploiting two CPUs using the parallel map function improves performance over the sequential map by 70% in this case:

```
> time (map fib) [|40; 40|];;
Took 3138ms
```

```
val it : int array = [|102334155; 102334155|]
```

However, applying this naive parallel map to a long array of quick computations increases the relative cost of this function's overheads. Consequently, the parallel map function can also be much slower than a sequential map, at least four orders of magnitude slower when mapping single increments over each element of a 10^4-element array:

```
> time (Array.map (( + ) 1)) [|1 .. 10000|];;
Took 0ms
val it : unit = ()
> time (map (( + ) 1)) [|1 .. 10000|];;
Took 9797ms
val it : unit = ()
```

Using a smaller number of threads and synchronizing their access to the input array greatly improves the worst-case performance of this parallel map. Implementing this improved parallel map requires the use of .NET thread synchronization constructs.

9.3.3 Locks

Two of the deficiencies of the naive parallel map can be addressed by distributing the element-wise computations over a small number of threads. Using fewer threads reduces the overheads of thread creation and handling and greatly improves worst-case performance, broadening the utility of the parallel map function.

A lock is as a way to mutually exclude threads from performing certain tasks concurrently, such as incrementing a counter. Exactly this functionality can be used to synchronise a small number of threads to process array elements concurrently, each thread consuming the next available element until no elements remain.

A section of code is most elegantly locked using the higher-order function `lock`:

```
> lock;;
val it : obj -> (unit -> 'a) -> 'a
```

This function waits until the given object is unlocked, then locks it, executes the given function and unlocks the object before returning. Thus, a thread safe increment of an `int ref` called n may be written:

```
lock n (fun () -> incr n)
```

This leads to a more efficient implementation of the parallel map that uses a small number of threads, each of which use a shared counter to keep track of the next unmapped element in the array. The task of incrementing the counter and returning the next unmapped element (if any) may be written:

```
> let next i n () =
    if !i = n then None else
      incr i
      Some(!i - 1);;
```

```
val next : int ref -> int -> unit -> int option
```

The parallel map itself spawns threads executing a loop function which repeatedly maps a single element and looks for the next unmapped element:

```
> let map max_threads f a =
    let n = Array.length a
    let b = Array.create n None
    let i = ref 0
    let rec apply i =
      b.[i] <- Some(f a.[i])
      loop()
    and loop() =
      Option.iter apply (lock i (next i n))
    execute max_threads loop
    Array.map Option.get b;;
val map : int -> ('a -> 'b) -> 'a array -> 'b array
```

For CPU-bound operations, this map should be invoked with one thread for each available CPU or core:

```
> let cpu_map f a =
    map System.Environment.ProcessorCount f a;;
val cpu_map : ('a -> 'b) -> 'a array -> 'b array
```

In the worse case described above, this implementation is three orders of magnitude faster than the previous implementation:

```
> time (cpu_map (( + ) 1)) [|1 .. 1000|] |> ignore;;
Took 5ms
val it : unit = ()
```

This cpu_map function is quite efficient and distributes computation as evenly as possible whilst remaining safe. Consequently, this is the parallel map of choice for scientific applications that value simplicity over performance. The overheads of this parallel map implementation are primarily the creation of local threads and the synchronization between those threads.

9.3.4 The thread pool

The previous implementation of parallel map distributed computations over its own set of threads. A set of threads that serves this purpose is known as a thread pool and the .NET platform actually provides a global thread pool that typically contains 25 worker threads and is accessed via the ThreadPool namespace.

The advantage of using a global thread pool is that the worker threads have already been created, removing this overhead from the parallel map function. The disadvantage is lack of safety: functions that use the threadpool recursively can deadlock threads in the threadpool and the global threadpool has substantial overheads for queueing so it is typically much slower to use.

The `QueueUserWorkItem` method can be used to queue computations on the global thread pool directly. Asynchronous delegates provide an alternative, and often easier, way to execute computations concurrently using the threadpool.

9.3.5 Asynchronous delegates

A delegate is the .NET representation of a type-safe function pointer. In F#, a closure can be converted into a delegate using the `System.Converter` class.

Delegates provide asynchronous invocation via `BeginInvoke` and `EndInvoke` methods. The former queues the delegate in the thread pool, applying any arguments, and the latter waits for it to complete and recovers its return value.

A parallel map that uses the global thread pool by invoking asynchronous delegates may be written:

```
> let global_map f a =
    let d = new System.Converter<'a, 'b>(f)
    Array.map (fun x -> d.BeginInvoke(x, null, null)) a
    |> Array.map (fun a -> d.EndInvoke(a));;
val global_map :
  int -> ('a -> 'b) -> 'a array -> 'b array
```

This implementation of a parallel map is faster than the previous implementation when the total time taken to perform the whole map is small (< 1ms) because the time taken to execute the previous map implementation is dominated by the thread creation. However, this implementation is significantly slower in almost all other cases because it queueing jobs for the threadpool is slower and there are many more threads in the global threadpool than CPUs so the computations are constantly context switched between.

Perhaps more importantly, this `global_map` function handles exceptions by reraising them on the main thread whereas the previous `cpu_map` function will give undefined behaviour because it makes no attempt to handle exceptions.

9.3.6 Background threads

On the .NET platform, applications have background threads and foreground threads. The difference between background and foreground threads relates to the termination of the application. An application is terminated when all of its foreground threads complete, i.e. any outstanding background threads are terminated automatically.

The `IsBackground` property can be used to mark a thread as a background thread:

```
> thread.IsBackground <- true;;
```

This allows branch computations to be quit automatically and is of particular importance in the context of visualization, where computations should be performed in background threads to ensure that the application ends when the user interface is closed.

9.4   RANDOM NUMBERS

The `System.Random` class provided by .NET can be used to generate uniformly-distributed `int` and `float` random numbers. A friendlier interface can be obtained by creating a global random number generator and providing some useful functions to use it:

```
> module Random =
    let rand = new System.Random()
    let int n = rand.Next(n)
    let float x = x * rand.NextDouble();;
```

The `int` and `float` functions in this `Random` module generate uniformly-distributed random `int` and `float` values, respectively. The results lie in the range $[0, x)$ where x is the function argument.

For example, random floats uniformly distributed between 0 and 3 may be generated using:

```
> Random.float 3.0;;
val it : float = 1.639560843
```

This `Random` module is an OCaml-compatible replacement of the `System.Random` class and will be used in the remainder of this book.

9.5   REGULAR EXPRESSIONS

Particularly with the advent of bioinformatics, a growing number of scientific applications are manipulating strings as well as numbers. The F# string type provides simple access to Unicode strings and can be used to perform a variety of simple actions. However, many applications benefit from sophisticated and specialized forms of pattern matching that are designed to act upon strings.

We have already discussed the concept of regular expressions in the context of the `fslex` lexer generator, in section 5.5.1.

Definitions relating to .NET regular expressions are held in the namespace:

```
> open System.Text.RegularExpressions;;
```

We shall now examine some of the functionality provided by a simple regular expression before presenting a complete program to compute the frequencies of different words in a text document.

The `Regex` class can be used to create an object representing a given regular expression. For example, the following regular expression matches any sequence of one or more whitespace characters:

```
> let whitespace = new Regex(@"\s+");;
val whitespace : Regex
```

Note that we have used the notation `@"..."` to automate the escaping of back-slashes in a string. The alternative would be to omit the @ and escape the strings manually, which is more tedious and error-prone: `"\\s+"`. This will be more beneficial for longer, more complicated regular expressions.

This .NET object of type `Regex` has several member functions that can be used to manipulate strings by finding substrings that match this regular expression.

For example, the `Replace` member function finds all substrings that match the regular expression and replaces them with the given string. The `Replace` member function of the `whitespace` object can therefore be used to collapse whitespace in a string by replacing all sequences of one or more whitespace characters with a single space:

```
> let collapse_whitespace string =
    whitespace.Replace(string, " ");;
val collapse_whitespace : string -> string
```

Applying this function to a string containing superfluous whitespace produces a new string with sequences of whitespace replaced by single spaces:

```
> collapse_whitespace "Too   many\n  spaces.";;
val it : string = "Too many spaces."
```

The same `is_whitespace` object can be used to split a string into whitespace-separated substrings using the `Split` member:

```
> whitespace.Split("Too many spaces.");;
val it : string = [|"Too"; "many"; "spaces."|]
```

This simple example has illustrated the basic functionality of .NET regular expressions. This functionality can be used to dissect simple file formats to interpret textual data.

9.6   VECTORS AND MATRICES

The F# standard library provides efficient functions for handling arbitrary-dimensionality vectors and matrices, including specialized versions for vectors and matrices with `float` elements. Vectors are considered to be column vectors by default. Row vectors are also supported and are distinguished from column vectors by the type system, to improve correctness.

Vectors and matrices of floats can be constructed from sequences and sequences of sequences using the `vector` and `matrix` functions, respectively:

```
> let r = vector [2.0; 3.0];;
val e : Math.vector
> let m = matrix [[0.0; 1.0]; [-1.0; 0.0]];;
val m : Math.matrix
```

Arithmetic operators are provided for vectors and matrices. Vector addition:

```
> r + r;;
val it : vector = vector [4.0; 6.0]
```

Scaling:

```
> 3.0 $* r;;
val it : vector = vector [6.0; 9.0]
```

Matrix multiplication:

```
> m * m;;
val it : Matrix<float> =
  matrix [[-1.0; 0.0]; [0.0; -1.0]]
```

Transformation of a column vector by premultiplying a matrix:

```
> m * r;;
val it : vector = vector [3.0; -2.0]
```

Transformation of a row vector by postmultiplying a matrix:

```
> Math.Vector.Transpose r * m;;
val it : vector = vector [-3.0; 2.0]
```

Many scientific problems can be phrased in terms of vector-matrix algebra. Thus, the ability to handle vectors and matrices can be instrumental in writing scientific programs. In particular, the ability to perform some complicated computations on them (e.g. finding the eigenvalues of a matrix) can be pivotal in scientific programs. Such computations are often prone to numerical error and, therefore, can be tedious to program robustly. Interfacing to existing libraries that implement this functionality is covered later in this chapter.

9.7   DOWNLOADING FROM THE WEB

The Worldwide Web (**WWW**) contains a wealth of information for scientists, much of it in a form that can be downloaded and analyzed by machine. Indeed, even the structure of the Web has been the subject of scientific research. Naturally, the .NET platform provides all of the tools required to tap this resource quickly and easily.

Definitions relating to networking and the internet are held in the System.Net class and definitions relating to IO are held in the System.IO namespace. The .NET libraries provide access to the Web such that programs can download files over an internet connection. A file to download a given URL by passing a stream to a function k may be written in terms of these classes:

```
> let download (url : string) k =
    let request = System.Net.WebRequest.Create(url)
    let response = request.GetResponse()
    use stream = response.GetResponseStream()
    k stream;;
val download : string -> (IO.Stream -> 'a) -> 'a
```

The function k is given the open stream and is expected to read everything it needs from the stream before returning, as the stream will be closed when the continuation returns.

The simplest continuation simply reads the stream into a string:

```
> let string_of_stream (stream : #System.IO.Stream) =
    (new System.IO.StreamReader(stream)).ReadToEnd();;
val string_of_stream : #IO.Stream -> string
```

For example, we can download the Google home page with:

```
> download "http://www.google.com" string_of_stream;;
val it : string
  = "<html><head><meta http-equiv=\"content-type...
```

This makes F# and, in particular, the F# interactive mode a powerful tool for scientists analyzing data available on the web.

9.8 COMPRESSION

The .NET standard library includes functions for compressing and uncompressing streams. Many on-line resources store information in compressed form and these functions can be used to uncompress the data ready for analysis.

The `System.IO.Compression` namespace contains definitions relating to data compression. The following `gunzip` function can be used to decompress a stream representing compressed data stored in the GZip format (commonly found on Unix systems):

```
> open System;;
> let gunzip (stream : #IO.Stream) =
    let mode = IO.Compression.CompressionMode.Decompress
    new IO.Compression.GZipStream(stream, mode);;
val gunzip : #IO.Stream -> IO.Compression.GZipStream
```

This function is used to decompress downloaded data in future example programs.

9.9 HANDLING XML

The .NET standard library includes many data structures and algorithms for manipulating data stored in the XML format. On-line scientific databases are increasingly using the XML format.

Definitions relating to XML are found in the `System.Xml` class.

9.9.1 Reading

Data in the XML format can be read from a stream, such as the streams generated by the `download` and `gunzip` functions defined above, into an object of the class `XmlDocument` by constructing such an object and invoking its `Load` method on the stream.

For example, a function to load an XML document from a stream (such as a GZip stream) may be written:

```
> open System;;
> let xml_of_stream stream =
    let doc = new Xml.XmlDocument()
    doc.Load(stream :> IO.Stream)
    doc;;
val xml_of_stream : #IO.Stream -> Xml.XmlDocument
```

We shall use this function in future examples dealing with XML.

9.9.2 Writing

Once created, an XML document `doc` can be written to a stream in XML format simply by invoking the `Save` method of the `doc` object. For example, the following prints `doc` to the console in XML format:

```
doc.Save(stdout);;
```

The ability to load and save XML documents using the .NET libraries is particularly useful given the increasing amount of scientific data found on the web in XML format. However, the F# programming language has many features that make it ideally suited to tree manipulation, including the manipulation of XML data. In order to leverage these language features, it is useful to translate the .NET representation of XML into a native F# variant type.

9.9.3 Declarative representation

The object-oriented approach used by the `System.Xml` library is not ideal for visualizing in the F# interactive mode or dissecting using pattern matching. Consequently, it is often beneficial to rewrite XML into a declarative form. A variant type can be used to represent XML data in a more F#-friendly way:

```
> type xml =
    | Element of string *
                 (string * string) list *
                 xml list
    | Text of string;;
```

For example, the XML snippet:

```
<doc name="My document">
  <title>This is a document</title>
</doc>
```

will be represented by the F# value:

```
Element("doc",
```

```
            ["name", "My document"],
            [Element("title",
                    [],
                    [Text "This is a document"])])])
```

Values of the XmlNode and XmlElement types can be converted to the F# variant type xml using the following pair of mutually-recursive functions:

```
> let rec node (n : Xml.XmlNode) = match n.NodeType with
    | Xml.XmlNodeType.Element ->
        element (n :?> Xml.XmlElement)
    | Xml.XmlNodeType.Text -> Text n.InnerText
    | _ -> invalid_arg "node"
  and element elt =
    Element(elt.LocalName,
            [ for attrib in elt.Attributes ->
                attrib.Name, attrib.Value ],
            [ for child in elt.ChildNodes ->
                node child ]);;
val node : Xml.XmlNode -> xml
val element : Xml.XmlElement -> xml
```

Note the use of the downcast operator :?> to convert a node from the parent XmlNode class to the derived XmlElement class.

An XML document doc of the type XmlDocument can be converted into a value of the type xml by applying this element function to its DocumentElement property:

```
element doc.DocumentElement
```

The declarative representation of an XML document as a value of the F# type xml, rather than an inheritance hierarchy of objects, allows functions to dissect XML data using pattern matching. This concept is taken a step further in chapter 10 by using active patterns to dissect an object-oriented representation directly without having to copy it into an F# variant type.

9.10   CALLING NATIVE LIBRARIES

The main advantage of native-code libraries is that they can provide much better performance than .NET libraries. However, this performance comes at a grave cost in terms of safety. Native-code libraries can cause all-manner of problems when used incorrectly, including random crashing. Consequently, native code libraries should be used only when their functionality is not available via a safer route (i.e. a .NET library) or when performance is critical. In the interests of correctness, the amount of unsafe code should be minimized and thoroughly checked before use.

A binding is a piece of code that describes how an external function or program can be invoked from within the host language (F#). In this case, bindings are F#

libraries that include references to external native code libraries. The simplest form of binding details the name *library* of the DLL, the function *func*$_1$ inside the DLL and the C signature of the function including its name *func*$_2$ in F#:

```
[<DllImport(@"library.dll", EntryPoint="func1")>]
extern double *func2(args);
```

Note that the `System.Runtime.InteropServices` namespace contains the definition of `DllImport`.

For example, the following F# snippet links to the library *mylib.dll*, binding the function `fib` to an F# function called `ext_fib` which has a signature `int -> int`:

```
[<DllImport(@"mylib.dll", EntryPoint="fib")>]
extern int ext_fib(int n);
```

For detailed information about writing bindings to native-code libraries, refer to the F# manual.

9.11 FOURIER TRANSFORM

The ability to compute a numerical approximation to the Fourier transform of a signal is of fundamental importance in scientific computing. A great deal of computer science research has been directed at this area and the latest Fast Fourier Transform (**FFT**) algorithms are capable of transforming n uniform samplings of a signal into Fourier space in $O(n \log n)$ time and with $O(\sqrt{\log n})$ mean error.

The FFT algorithm is based upon non-trivial results from number theory. Specifically, the logarithm in the complexity of the FFT stems from the recursive subdivision of the n-element input vector into $\frac{n}{p}$-element subvectors where p is a small prime factor of n. When n does not have any small prime factors (e.g. when n itself is prime), the algorithm expands the input into a larger m-element vector where $m > 2n$ such that m has many small prime factors and can then be subdivided efficiently. Although the performance follows an $n \log n$ trend, the results for individual n are strongly dependent upon the factorization of n and, consequently, appear "noisy". The variation in performance between consecutive n can be as much as a factor of four and, consequently, larger values of n can actually be significantly faster.

9.11.1 Native-code bindings

The FFTW library, developed at MIT, is the best freely-available FFT implementation. This section details minimal bindings to the FFTW library.

The Fourier transform and inverse Fourier transform are referred to as forward and backward transforms:

```
> type direction = Forward | Backward;;
```

Searching possible factorizations can take as long as the FFT itself but factorization plans can be reused when computing many FFTs of the same length. Two of search algorithms implemented by FFTW allow a factorization plan to be estimated from general performance information or calculated more accurately by making real performance measurements:

```
> type precalc = Estimate | Measure;;
```

In the interests of performance, FFTW aligns data to CPU-friendly boundaries and provides functions to allocate and free aligned memory:

```
> open System.Runtime.InteropServices;;
> [<DllImport(@"libfftw3-3.dll",
              EntryPoint="fftw_malloc")>]
  extern double *fftw_malloc(int size);;
val fftw_malloc : int -> double nativeptr

> [<DllImport(@"libfftw3-3.dll",
              EntryPoint="fftw_free")>]
  extern void fftw_free(double *data);;
val fftw_free : double nativeptr -> unit
```

Factorization plans are created by the `fftw_plan_dft_1d` function:

```
> [<DllImport(@"libfftw3-3.dll",
              EntryPoint="fftw_plan_dft_1d")>]
  extern void *plan_dft_1d_(int n, double *i, double *o,
                            int sign, int flags);;
val plan_dft_1d_ :
  int -> double nativeptr -> double nativeptr ->
    int -> int -> nativeint
```

Note that this function has been bound to the F# function `plan_dft_1d_` where the final underscore is taken to indicate an unsafe binding. The `i` and `o` arguments are the input and output vectors and the return value (declared to be of the C type `void *`) is an FFTW "plan".

The `sign` and `flag` arguments to the `plan_dft_1d_` function can be derived from `direction` and `precalc` values:

```
> let plan_dft_1d n i o d e =
    let d =
      match d with
      | Forward -> -1
      | Backward -> 1
    let flags =
      match e with
      | Estimate -> 64
      | Measure -> 0
    plan_dft_1d_ (n, i, o, d, flags + 1);;
val plan_dft_1d :
  int -> double nativeptr -> double nativeptr ->
```

```
direction -> precalc -> nativeint
```

Note that the `flags` variable always has bit 1 set, allowing FFTW to alter the input array.

The `execute` function is the only thread-safe function provided by FFTW and it computes the FFT for a given plan.

```
> [<DllImport(@"libfftw3-3.dll",
                EntryPoint="fftw_execute")>]
  extern void execute(void *plan);;
val execute : nativeint -> unit
```

9.11.2 Interface in F#

The following function creates an FFTW-compatible array to store n complex numbers, returning both the native pointer to the array and the `NativeArray` itself:

```
> let make n =
    let ptr = fftw_malloc(16 * n)
    let na = NativeArray.FromPtr(ptr, n)
    ptr, na;;
val make : int -> double nativeptr * NativeArray<double>
```

Note that the argument to `fftw_malloc` is the size of the array in bytes. In this case there are 8 bytes per float and two floats per complex number.

The following function sets the ith element of the native array a to the complex number z:

```
> let set (a : NativeArray<double>) i (z : Complex) =
    a.[2*i] <- z.r
    a.[2*i + 1] <- z.i;;
val set : NativeArray<double> -> int -> Complex -> unit
```

The following function gets the ith element of the native array a as a complex number:

```
> let get (a : NativeArray<double>) i =
    Math.Complex.Create(a.[2*i], a.[2*i + 1]);;
val get : NativeArray<double> -> int -> Complex
```

Rather than deleting factorization plans, we shall memoize them indefinitely in a hash table for later reuse:

```
> let plan =
    memoize (fun (n, d, e) ->
      let ptr, array = make n
      a, plan_dft_1d n ptr ptr d e);;
val plan :
  (int -> direction -> precalc ->
```

```
NativeArray<double> * nativeint)
```

A Fourier transform can be computed in the given direction using a controllable amount of precomputation for the factorization by creating a plan, copying the vector into the plan, applying the execute function and copying the data back out:

```
> let fft direction effort a =
    let n = Array.length a
    if n=0 then [||] else
    let array, plan = plan n direction effort
    Array.iteri (set array) a
    execute plan
    let s = complex (1.0 / sqrt(float n)) 0.0
    Array.init n (fun i -> s * get array i);;
val fft :
  direction -> precalc -> Complex array -> Complex array
```

Finally, we can define some easy-to-use functions for computing the Fourier transforms of complex arrays:

```
> let fourier = fft Forward Estimate;;
val fourier : Complex array -> Complex array
> let ifourier = fft Backward Estimate;;
val fourier : Complex array -> Complex array
```

The library functions can then be used.

9.11.3 Pretty printing complex numbers

In the interests of clarity, let us define and register a pretty printer for the Complex type:

```
> let chop x =
    if abs x < sqrt epsilon_float then 0.0 else x;;
val chop : float -> float
> let string_of_complex (z : Complex) =
    match chop z.Real, chop z.Imag with
    | 0.0, 0.0 -> "0"
    | r, 0.0 -> sprintf "%g" r
    | 0.0, i -> sprintf "%gi" i
    | r, i when i < 0.0 -> sprintf "%g - %gi" r (-i)
    | r, i -> sprintf "%g + %gi" r i;;
val string_of_complex : Math.Complex -> string
> fsi.AddPrinter(string_of_complex);;
val it : unit = ()
```

Complex numbers are now chopped and pretty printed in a much more concise way.

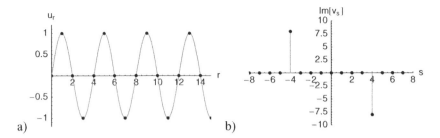

Figure 9.2   Fourier series of a discretely sampled sine wave, showing: a) the samples u_r $r \in [0, 16)$ and Fourier series $\sin(\frac{\pi}{2}r)$, and b) the corresponding Fourier coefficients v_s computed numerically using FFTW.

9.11.4 Example use

As an example, the following creates a 16-element array a containing four repeats of $(0, 1, 0, -1)$:

```
> let a =
    [| for i in 0 .. 15 ->
          complex (float i / 2.0 * Math.PI |> sin) 0.0 |]
val a : Complex array

> a;;
val it : Complex array =
    [|0; 1; 0; -1; 0; 1; 0; -1; 0; 1; 0; -1; 0; 1; 0; -1|]
```

This discrete sampling is illustrated in figure 9.2a. We shall now calculate the functional form of the Fourier series via the numerically computed series found using FFTW.

Taking the samples to be unit-separated samples, the Nyquist frequency is $\nu_{Ny} = \frac{1}{2}$. The signal a may be considered to be a sampling of a real-valued sinusoid $A\sin(2\pi\nu r)$ with amplitude $A = 1$ and frequency $\nu = \frac{1}{2}\nu_{Ny} = \frac{1}{4}$.

The FFT of a is:

```
> fourier a;;
val it : Complex [] =
    [|0; 0; 0; 0; -2i; 0; 0; 0; 0; 0; 0; 0; 2i; 0; 0; 0|]
```

Each element v_s of this array may be taken to represent the frequency $\nu(s) = s\nu_{Ny}/n$ and amplitude $A = \frac{1}{n}v_s$ of Fourier components in the signal. As we are dealing with Fourier series, the indices s are periodic over n. Consequently, we may productively interpret the second half of this result as representing negative frequencies, as illustrated in figure 9.2b.

In this case, the only non-zero elements are $v_4 = -8i$ and $v_{12} = 8i$. This shows that the signal can be represented as the sum of two plane waves which, in fact, partly cancel to give a sine wave:

$$
\begin{aligned}
f(r) &= \frac{1}{\sqrt{n}} \left(\sum_{0 \le s < \frac{1}{2}n} v_s e^{2\pi i r s/n} + \sum_{\frac{1}{2}n \le s < n} v_s e^{2\pi i r(s/n-1)} \right) \\
&= \frac{1}{\sqrt{n}} \left(v_4 e^{2\pi i r 4/16} - v_{12} i e^{2\pi i r(12/16-1)} \right) \\
&= \frac{1}{4} \left((-2i)i \sin(\frac{\pi}{2}r) + (2i)i \sin(-\frac{\pi}{2}r) \right) \\
&= \sin(\frac{\pi}{2}r)
\end{aligned}
$$

as expected.

Also, the inverse FFT of the FFT of a recovers the original a to within a numerical error of 10^{-16}:

```
> ifourier (fourier a);;
val it : Complex [] =
  [|0; 1; 0; -1; 0; 1; 0; -1; 0; 1; 0; -1; 0; 1; 0; -1|]
```

This FFT library allow the Fourier series of large data sets with up to three dimensions to be computed efficiently.

9.12   METAPROGRAMMING

The generation of explicit subprograms is a form of *metaprogramming*. The F# programming language makes metaprogramming easy for two reasons:

1. The F# language is derived from languages that were designed to manipulate programs.

2. The .NET platform provides light-weight code generation, allowing its intermediate language (**IL**) to be generated and invoked at run-time.

One of the most important aspects of metaprogramming is *partial specialization*. This refers to the generation of specialized subprograms that can perform a task more efficiently than a generic program.

Metaprogramming is currently available via the direct generation of IL code and via a high-level interface provided by the LINQ library in .NET 3.

9.12.1   Emitting IL code

Definitions relating to IL code generation are contained in the following namespaces:

```
> open System;;
> open System.Reflection;;
```

```
> open System.Reflection.Emit;;
```

A dynamically-generated function called f that accepts one int parameter and returns an int can be created with:

```
> let dm =
    new DynamicMethod("f",
                      (type int),
                      [|(type int)|],
                      (type obj));;
val dm : DynamicMethod
```

An object used to generate the IL code for this new function can be obtained with:

```
> let il = dm.GetILGenerator();;
val il : ILGenerator
```

IL code is a stack-based bytecode, meaning that the expression $(x + 2) \times 3$ is represented by commands to push argument zero (x) and the constant 2, followed by an add to consume the two stack entries and push the value of $x + 2$:

```
> il.Emit(OpCodes.Ldarg, 0);;
val it : unit = ()
> il.Emit(OpCodes.Ldc_I4, 2);;
val it : unit = ()
> il.Emit(OpCodes.Add);;
val it : unit = ()
```

Then push 3 and consume two stack entries with a multiply:

```
> il.Emit(OpCodes.Ldc_I4, 3);;
val it : unit = ()
> il.Emit(OpCodes.Mul);;
val it : unit = ()
```

Finally, return with the value on the top of the stack:

```
> il.Emit(OpCodes.Ret);;
val it : unit = ()
```

This completes the definition of a new function. The function can be invoked using the Invoke method and supplying an obj array of the arguments. For example, $3(x + 2) = 21$ for $x = 5$:

```
> (dm.Invoke(null, [|box 5|]) :?> int);;
val it : int = 21
```

Direct generation of IL code is harder and more error prone than using a high level library. However, the high-level libraries available at the time of writing lack some important features that may impede the use of metaprogramming in scientific computing.

9.12.2   Compiling with LINQ

The LINQ project is a new part of .NET that provides metaprogramming as a way to improve cross-language interoperability and, in particular, database querying. Amongst other things, this library provides a simple abstract syntax tree representation of expressions and a compiler that can convert these high-level abstract syntax trees into IL code.

At the time of writing, the LINQ project installs into its own directory:

```
> #I @"C:\Program Files\LINQ Preview\Bin";;
```

The following DLL contains the required functionality:

```
> #r "System.Query.dll";;
```

The following namespaces contain relevant definitions:

```
> open System;;
> open System.Expressions;;
```

The following .NET delegate type can be used to represent a function type for single-argument functions:

```
> type ('a, 'b) fn = delegate of 'a -> 'b;;
```

In the interests of clarity, we use the following abbreviation:

```
> type E = Expression;;
```

The following `int` function creates a constant integer expression:

```
> let int (n : int) = E.Constant(n);;
val int : int -> ConstantExpression
```

The two following binary operators `+:` and `*:` are used to compose expressions representing arithmetic operations:

```
> let ( +: ) f g = E.Add(f, g);;
val ( +: ) :
  #Expression -> #Expression -> BinaryExpression
> let ( *: ) f g = E.Multiply(f, g);;
val ( *: ) :
  #Expression -> #Expression -> BinaryExpression
```

A compiled delegate representing the λ-expression $x \to (x + 2) \times 3$ may be generated with:

```
> let f : fn<int, int> =
    let x = E.Parameter((type Int32), "x")
    (E.Lambda((x +: int 2) *: int 3, x)).Compile();;
val f : (int, int) fn
```

Note the type annotation for `f` is required for this example to work.

The generated delegate function f can be invoked with its argument using the Invoke method with

```
> f.Invoke(5);;
val it : int = 21
```

Metaprogramming can be used in scientific computing to improve the performance of many algorithms such as low-dimensional vector and matrix manipulation and wavelet/Fourier transforms.

CHAPTER 10

DATABASES

An increasing amount of scientific data, information and literature is available on the internet in the form of scientific databases. Unlike their industrial counterparts, on-line scientific databases are typically exposed either as repositories of XML data or in the form of remote procedure calls (**RPC**s) that can be invoked over the internet as web services.

Of the scientific disciplines, the life sciences are currently by far the most advanced in terms of on-line databases. This is primarily due to the explosion in bioinformatics-related data, most notably DNA and protein sequences, that may now be interrogated over the internet.

The .NET framework was specifically designed to cater for the requirements of on-line databases, XML data and web services. Consequently, on-line scientific databases can be accessed quickly and efficiently from F# programs by leveraging the .NET framework. This chapter describes the use of existing .NET functionality to interrogate two of the most important scientific databases in the life sciences: the Protein Data Bank (**PDB**) and GenBank.

F# for Scientists. By Jon Harrop
Copyright © 2008 John Wiley & Sons, Inc.

10.1   PROTEIN DATA BANK

The PDB provides a variety of tools and resources for studying the structures of biological macromolecules and their relationships to sequence, function, and disease. The PDB is maintained by the Research Collaboratory for Structural Bioinformatics (**RCSB**), a non-profit consortium dedicated to improving the understanding of biological systems function through the study of the 3-D structure of biological macromolecules.

10.1.1   Interrogating the PDB

The PDB provides a simple web interface that allows compressed XML files containing the data about a given protein to be downloaded over the web. This section describes how the XML data for a given protein may be downloaded and uncompressed using F# functions that are designed to be as generic as reusable as possible.

Each protein in the PDB has a four character identifier. The following function converts such an identifier into the explicit URL of the compressed XML data:

```
> let pdb name =
    "ftp://ftp.rcsb.org" +
    "/pub/pdb/data/structures/divided/XML/" +
    String.sub name 1 2 + "/" + name + ".xml.gz";;
val pdb : string -> string
```

The XML data are compressed in the GZip format. So the XML data for a given protein may be downloaded and uncompressed by combining the `pdb` and `gunzip` (see section 9.8) and `xml_of_stream` (see section 9.9) functions into a single `download` function:

```
> let download protein =
    let url = pdb protein
    let request = System.Net.WebRequest.Create(url)
    let response = request.GetResponse()
    use stream = response.GetResponseStream()
    gunzip stream
    |> xml_of_stream;;
val download : string -> System.Xml.XmlDocument
```

This `download` function can be used to obtain the XML data for an individual protein:

```
> let doc = download "1hxn";;
val doc : System.Xml.XmlDocument
```

The resulting value is stored in the object-oriented representation of XML data used by the .NET platform, referred to as the `System.Xml.XmlDocument` type. This native .NET data structure allows XML data to be constructed and dissected using object-oriented idioms.

The default printing of XML data in F# interactive sessions as nested sequences is of little use:

```
> doc;;
val it : System.Xml.XmlDocument
= seq
    [seq [];
     seq
       [seq ...
```

Consequently, an pretty printer for XML data is very useful when dissecting XML data interactively using F#.

10.1.2   Pretty printing XML in F# interactive sessions

The XmlDocument type represents a whole XML document and includes global metainformation about the data itself and the method of encoding. XML data is essentially just a tree and, ideally, we would like to be able to visualize this tree in interactive sessions and dissect it using pattern matching.

Nodes in an XML tree are derived from the base type XmlNode. The two most important kinds of node are Element and Text nodes. In XML, an Element node corresponds to a tag such as <br /> and a Text node corresponds to verbatim text such as "Title" in <h1>Title</h1>.

As described in section 2.6.4.3, pretty printers can be installed in a running interactive session to improve the visualization of values of a certain type. In this case, we require a pretty printer for XML documents and nodes:

```
> fsi.AddPrinter(fun (xml : System.Xml.XmlNode) ->
    xml.OuterXml);;
```

Viewing an XML tree in the interactive session now displays the string representation of the XML:

```
> doc;;
val it : System.Xml.XmlDocument =
<?xml version="1.0" encoding="UTF-8"?><PDBx:...
```

This pretty printer makes it much easier to dissect XML data interactively using F#.

10.1.3   Deconstructing XML using active patterns

XML data could be examined elegantly using pattern matching by translating a .NET XML tree of objects into an F# variant type (as discussed in section 9.9) and then dissecting the tree using pattern matching. However, the creation of an intermediate tree is wasteful.

A much more efficient way to dissect XML data is to use one of the unique features of F# called active patterns or *views*. Active patterns allow a non-F# data structure, such as the object-oriented .NET representation of an XML node, to be viewed as if it were a native F# variant type. This is achieved by defining active pattern constructors (in this case called `Element` and `Text`) that present the .NET data structure in a form suitable for pattern matching.

Before defining the active pattern, we define an `element` function that extracts the tag name, attributes and child nodes from an `Element` node:

```
> let element (elt : System.Xml.XmlElement) =
    elt.LocalName,
    seq { for attrib in elt.Attributes ->
            attrib.Name, attrib.Value },
    seq { for child in elt.ChildNodes ->
            child };;
val element :
  System.Xml.XmlElement ->
    string * seq<string * string> *
    seq<System.Xml.XmlNode>
```

The following defines two active patterns called `Element` and `Text` that match the respective kinds of XML node using the `:?` construct to detect the run-time type of an object:

```
> let (|Element|Text|) (n : System.Xml.XmlNode) =
    match n with
    | :? System.Xml.XmlElement as n ->
        Element(element n)
    | n -> Text n.InnerText;;
val ( |Element|Text| ) :
  System.Xml.XmlNode ->
  Choice<(string * seq<string * string> *
          seq<System.Xml.XmlNode>),string>
```

The following `acids_of_xml` function uses the active pattern in sequence comprehensions to extract the amino-acid sequence from the given XML document for a protein:

```
> let acids_of_xml (doc : System.Xml.XmlDocument) =
    let cs = doc.DocumentElement.ChildNodes
    seq { for Element("entity_poly_seqCategory", _, cs)
              in cs
            for Element("entity_poly_seq", attribs, _)
                in cs
              for "mon_id", acid in attribs ->
                acid };;
val acids_of_protein : System.Xml.XmlNode -> seq<string>
```

This is a beautiful example of the power and flexibility of active patterns and the comprehension syntax. The lines of this function represent progressively more fine-grained dissections of the XML document. The `entity_poly_seqCategory` tag is extracted first followed by the nested `entity_poly_seq` tag and, finally, the attribute called `mon_id` that contains the amino acid sequence. By wrapping the whole function in a single sequence comprehension syntax { . . . }, all of the matching attributes in the nested tag are listed sequentially with minimal effort.

The function can be used in combination with the previous functions to extract the amino-acid sequence of a given PDB protein directly from the on-line database:

```
> acids_of_xml doc;;
val it : seq<string> =
  seq ["HIS"; "ARG"; "ASN"; "SER"; ...]
```

This makes F# an extraordinarily powerful tool for the interactive analysis of PDB data. However, this example would not be complete without describing how easily the data structures involved can be visualized in a GUI to make data dissection easier than ever before.

10.1.4 Visualization in a GUI

The tree-based representation of XML data is ideally suited to graphical visualization using the ubiquitous `TreeView` Windows Forms control. This is the same control used to provide the default tree representation of disc storage on the left hand side of Windows Explorer.

Windows Forms programming makes heavy use of the following namespace:

```
> open System.Windows.Forms;;
```

We begin by creating a new Windows form:

```
> let form =
    new Form(Visible=true, Text="Protein data");;
val form : Form
> form.TopMost <- true;;
val form : Form
```

and adding an empty `TreeView` control to it, making sure that the control expands to fill the whole window:

```
> let tree = new TreeView(Dock=DockStyle.Fill);;
val tree : TreeView
> form.Controls.Add(tree);;
val it : unit = ()
```

The following function traverses the tree representation of an XML data structure using the active patterns `Element` and `Text` defined above, accumulating a `TreeNode` ready for insertion into the `TreeView`:

```
> let rec treeview_of_xml = function
    | Text string -> new TreeNode(string)
    | Element(tag, attribs, cs) ->
```

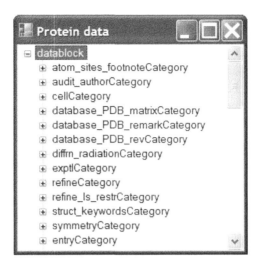

Figure 10.1   Using a Windows Forms `TreeView` control to visualize the contents of an XML document from the Protein Data Bank.

```
        let parent = new TreeNode(tag)
        for n in cs do
          parent.Nodes.Add(treeview_of_xml n) |> ignore
        parent;;
val treeview_of_xml : XmlNode -> TreeNode

> (root :> System.Xml.XmlNode)
  |> treeview_of_xml
  |> tree.Nodes.Add;;
val it : unit = ()
```

The result is shown in figure 10.1. Even a minimal GUI such as this can be instrumental in interactive data dissection, greatly accelerating the development process.

10.2 WEB SERVICES

The term "web services" refers to any service provided on-line by a remote computer. Typically, a web service is a programmatic way to interrogate the contents of a database held on the remote server. In some cases, web services are used to combine and process information from various sources.

The Simple Object Access Protocol (**SOAP**) is by far the most popular way to access web services. A SOAP API consists of a variety of dynamically-typed functions known as remote procedure calls (**RPC**s) that can be used to request information from a web service. SOAP APIs are typically encapsulated in a single definition using a format known as the Web Service Definition Language (**WSDL**). A WSDL description of a SOAP API may be automatically compiled into statically-typed function definitions, making it easier to avoid type errors when using web services.

The use of web services in F# revolves around the use of web references in C#, creating a C# DLL and linking to it from an F# program. The process of creating C# DLLs and referencing them from F# code was described in section 2.6.5.

To add a web reference to a C# DLL project, right click on "References" in Solution Explorer and select "Add Web Reference...". Paste the URL of the WSDL file into this window and click "Go". Once the WSDL file has been downloaded and examined, Visual Studio gives an overview of the API described by the file. Adding a WSDL web reference to a project causes Visual Studio to autogenerate thousands of lines of C# source code implementing the whole of the API provided by that web service (the description of which was in the WSDL file). By compiling the autogenerated C# code into a .NET DLL and linking to it from F#, web services can be used from F#.

10.2.1   US temperature by zip code

As a very simple initial example, create a Visual Studio solution composed of a C# DLL and an F# program referencing the DLL (following the description in section 2.6.5) and add a web reference to the URL:

```
http://www.xmethods.com/sd/TemperatureService.wsdl
```

Now build the C# DLL and use the #r directive in the F# code to load the C# DLL and then open its namespace:

```
> #r "ClassLibrary1.dll";;
> open ClassLibrary1;;
```

This web service simply provides a getTemp function that returns the current temperature in Fahrenheit as a float in the region of a US zipcode given as a string.

A new instance of the TemperatureService class must be created in order to use this web service:

```
> let server =
    new net.xmethods.www.TemperatureService();;
val server : TemperatureService
```

This class provides the getTemp function as a member. Invoking this member function causes the remote procedure call to be made to the SOAP web service hosted at xmethods.com and should return the temperature in Beverly Hills in the following case:

```
> server.getTemp("90210");;
val it : float32 = 52.0f
```

This is a minimal example demonstrating the creation of a C# DLL, the use of web references and the interoperability between F# and C# to use some of the C# tools from F# programs.

10.2.2   Interrogating the NCBI

The National Center for Biotechnology Information (**NCBI**) was established in 1988 as a national resource for molecular biology information. The NCBI creates public databases, conducts research in computational biology, develops software tools for analyzing genome data, and disseminates biomedical information.

The NCBI web service enables developers to access Entrez Utilities using SOAP. Programmers may write software applications that access the E-Utilities in F#. The WSDL file describing the E-Utilities SOAP interface is available at the following URL:

```
http://www.ncbi.nlm.nih.gov/entrez/eutils/soap/eutils.wsdl
```

This web reference can be made accessible from F# by following the same procedure as before and creating an instance of the eUtilsService class:

```
> let serv =
    new gov.nih.nlm.ncbi.eutils.eUtilsService();;
val serv : gov.nih.nlm.ncbi.eutils.eUtilsService
```

Member functions of this serv object with names of the form run_e*_MS may be used to interrogate the NCBI databases in a variety of ways from the Microsoft platform. The NCBI web service is more comprehensive and correspondingly more complicated than the previous examples. These functions essentially ask the database a question and the object returned is loaded with many different forms of metadata, including the parameters of the call itself, as well as the answer to the question.

The run_eInfo_MS member function of the serv object can be used to obtain general information regarding the web service:

```
> let res = serv.run_eInfo_MS("", "", "");;
val res : gov.nih.nlm.ncbi.eutils.eInfoResultType
```

In particular, the DbList property of the result res contains a list of the databases available:

```
> res.DbList;;
val it : gov.nih.nlm.ncbi.eutils.DbListType
= CSWebServiceClient.gov.nih.nlm.ncbi.eutils.DbListType
    {Items = [|"pubmed"; "protein"; "nucleotide";
              "nuccore"; "nucgss"; "nucest";
              "structure"; "genome"; "books";
              "cancerchromosomes"; "cdd"; "gap";
              "domains"; "gene"; "genomeprj"; "gensat";
              "geo"; "gds"; "homologene"; "journals";
              "mesh"; "ncbisearch"; "nlmcatalog";
              "omia"; "omim"; "pmc"; "popset";
              "probe"; "proteinclusters"; "pcassay";
              "pccompound"; "pcsubstance"; "snp";
              "taxonomy"; "toolkit"; "unigene";
```

```
                "unists"|];}
```

This list of databases spans a range of different topics within molecular biology, including literature (pubmed), proteins and DNA sequences (nucleotide).

10.2.2.1   ESearch   The web service for the ESearch utility allows users to perform searches on the databases to find database records that satisfy given criteria.

For example, the *Protein* database can be searched for proteins that have a molecular weight of $200, 020$:

```
> let res =
    serv.run_eSearch_MS("protein",
                        "200020[molecular+weight]",
                        "", "", "", "", "", "", "", "",
                        "", "", "0", "15", "", "");;
val it : eutils.eSearchResultType
```

The result `res` is an object that encapsulates the IDs of the four proteins found by the search:

```
> res.IdList;;
val it : string array =
  [|"16766766"; "4758956"; "16422035"; "4104812"|]
```

The `res` object also carries a variety of meta information including the parameters of the search:

```
> res;;
val it : gov.nih.nlm.ncbi.eutils.eSearchResultType =
  gov.nih.nlm.ncbi.eutils.eSearchResultType
    {Count = "4";
     ERROR = null;
     ErrorList = null;
     IdList = [|"16766766"; "4758956"; "16422035";
                "4104812"|];
     QueryKey = null;
     QueryTranslation = "000200020[molecular weight]";
     RetMax = "4";
     RetStart = "0";
     TranslationSet = [||];
     TranslationStack =
       [|gov.nih.nlm.ncbi.eutils.TermSetType;
         "GROUP"|];
     WarningList = null;
     WebEnv = null;}
```

The ESearch utility is the most general purpose way to interrogate the database and, consequently, it is the most useful approach. For example, repeated searches might be used to determine the distribution of the molecular weights of proteins in the database for different kinds of protein.

10.2.2.2   EFetch utility   The web service to the EFetch utility retrieves specific records from the given NCBI database in the requested format according to a list of one or more unique identifiers.

For example, fetching the record from the taxonomy database that has the ID 9, 685:

```
> let res = serv.run_eFetch_MS("taxonomy", "9685",
                               "", "", "", "", "", "",
                               "", "", "", "", "", "");;
val res : gov.nih.nlm.ncbi.eutils.eFetchResultType
```

Once again, the result `res` encapsulates information about the database record that was fetched. In this case, the following three properties of the `res` object describe the result:

```
> res.TaxaSet.[0].ScientificName;;
val it : string = "Felis catus"
> res.TaxaSet.[0].Division;;
val it : string = "Mammals"
> res.TaxaSet.[0].Rank;;
val it : string = "species"
```

These properties of the response `res` show that this record in the taxonomy database refers to cats.

As the EFetch utility fetches specific records, its use is much more limited than the ESearch utility. However, fetching specific database records is much faster than searching the whole database for all records satifying given criteria.

10.3   RELATIONAL DATABASES

Thus far, this chapter has described how F# programs can interrogate a variety of third party repositories of information in order to mine them for relevant data. Many of these repositories are actually stored in the form of relational databases exposed via web services. Relational database technology has many advantages such as simplifying and optimizing searches and, in particular, handling concurrent reads and writes by many different users or programs. This technology is ubiquitous in industry and is used to store everything from interactive data from company websites to client databases. Consequently, Microsoft have developed one of the world's most advanced database systems available, in the form of SQL Server. Moreover, this software is freely available in a largely-unrestricted form. So SQL Server can be a valuable tool for scientists wanting to maintain their own repositories of information.

This section describes how instances of SQL Server can be controlled from F# programs using Microsoft's ADO.NET interface. However, exactly the same approach can be used to manipulate and interrogate many other relational database implementations including Firebird.NET, MySql and Oracle. There is a vast body of literature with more detailed information on relational databases [25, 15].

The relational database interfaces for SQL Server are provided in the following two namespaces:

```
> open System.Data;;
> open System.Data.SqlClient;;
```

Before a database can be interrogated, a connection to it must be opened.

10.3.1 Connection to a database

Databases are connected to using "connection strings". Although these can be built naively using string concatenation, that is a potential security risk because incorrect information might be supplied by the user and injected into the database by the application. So programmatic construction of connection strings is more secure. The `SqlConnectionStringBuilder` class provides exactly this functionality:

```
> let connString = new SqlConnectionStringBuilder();;
val connString : SqlConnectionStringBuilder
```

The following lines indicate that we're accessing a database hosted by SQL Server Express using its integrated security feature (the connection might be rejected if we don't specify this):

```
> connString.DataSource <- @".\SQLEXPRESS";;
val it : unit = ()
> connString.IntegratedSecurity <- true;;
val it : unit = ()
```

The following lines create a `SqlConnection` object that will use our connection string when opening its connection to the database:

```
> let conn = new SqlConnection();;
val conn : SqlConnection
> conn.ConnectionString <- connString.ConnectionString;;
val it : unit = ()
```

The connection is actually opened by calling the `Open` member of the SQL connection:

```
> conn.Open();;
val it : unit = ()
```

An open database connection can be used to interrogate and manipulate the database.

10.3.2 Executing SQL statements

The *Simple Query Language* (**SQL**) is actually a complete programming language used to describe queries sent to databases such that they may be executed quickly and efficiently. Given that SQL is a text-based language, it is tempting to construct queries by concatenating strings and sending the result as a query. However, this is a bad idea because this approach is insecure[20]. Specifically, injecting parameters into a query by concatenating strings will fail if the parameter happens to be valid SQL and, worse, the parameter might contain SQL code that results in data being deleted or corrupted. The safe alternative is to construct SQL queries programmatically using the `SqlCommand` class.

The following function can be used to execute SQL statements:

```
> let execNonQuery conn s =
    let comm =
      new SqlCommand(s, conn, CommandTimeout=10)
    try
      comm.ExecuteNonQuery() |> ignore
    with e ->
      printf "Error: %A\n" e;;
val execNonQuery : SqlConnection -> string -> unit
```

For example, we can now create our database using the SQL statement CREATE DATABASE:

```
> execNonQuery conn "CREATE DATABASE chemicalelements";;
val it : unit = ()
```

Similarly, we can create a database table to hold the data about our chemical elements:

```
> execNonQuery conn "CREATE TABLE Elements (
Name varchar(50) NOT NULL,
Number int NOT NULL,
Weight float NOT NULL,
PRIMARY KEY (Number))";;
val it : unit = ()
```

The SQL type `varchar` (n) denotes a string with a maximum length of n. Note that we are using the atomic number as the primary key for this table because this value uniquely identifies a chemical element.

The following SQL statements add two rows to our database table, for Hydrogen and Helium:

```
> execNonQuery conn "INSERT INTO Elements
(Name, Number, Weight)
VALUES ('Hydrogen', 1, 1.008)";;
val it : unit = ()

> execNonQuery conn "INSERT INTO Elements (Name, Number,
Weight)
```

---

[20]Even if the database is private and will not be subjected to malicious attacks it can still be corrupted accidentally.

```
VALUES ('Helium', 2, 4.003)"
val it : unit = ()
```

In addition to manipulating a database using SQL statements, the contents of a database can be interrogated using SQL expressions.

10.3.3 Evaluating SQL expressions

We can query the database to see the current contents of our table using the following function:

```
> let query() =
    let query =
      "SELECT Name, Number, Weight FROM Elements"
    seq { let conn_string = connString.ConnectionString
          use conn = new SqlConnection(conn_string)
          do conn.Open()
          use comm = new SqlCommand(query, conn)
          use reader = comm.ExecuteReader()
          while reader.Read() do
            yield (reader.GetString 0,
                   reader.GetInt32 1,
                   reader.GetDouble 2) };;
val query : unit -> seq<string * int * float>
```

Note the use of an imperative sequence expression to enumerate the rows in the database table and dispose of the connection when enumeration of the sequence is complete. The reader object is used to obtain results for given database columns of the expected type. In this case, columns 0, 1 and 2 contain strings (chemical names), ints (atomic numbers) and double-precision floats (atomic weights).

Executing this query returns the data for Hydrogen and Helium from the database as expected:

```
> query();;
val it : seq<string * int * float> =
  seq [("Hydrogen", 1, 1.008); ("Helium", 2, 4.003)]
```

Accessing databases by encoding SQL commands as strings is fine for trivial, rare and interactive operations like creating the database itself but is not suitable for more sophisticated use. To perform significant amount of computation on the database, such as injecting the data we downloaded from the web service, we need to access the database programmatically.

10.3.4 Interrogating the database programmatically

The `SqlDataAdapter` acts as a bridge between a `DataSet` and SQL Server for retrieving and saving data:

```
> let dataAdapter = new SqlDataAdapter();;
val dataAdapter : SqlDataAdapter
```

A `DataSet` is an in-memory cache of data retrieved from a data source. The following function queries our database and fills a new `DataSet` with the results:

```
> let buildDataSet conn query =
    dataAdapter.SelectCommand <-
      new SqlCommand(query, conn)
    let dataSet = new DataSet()
    new SqlCommandBuilder(dataAdapter) |> ignore
    dataAdapter.Fill dataSet |> ignore
    dataSet;;
val buildDataSet : SqlConnection -> string -> DataSet
```

For example, the following query finds all rows in the database table `Elements` and returns all three columns in each:

```
> let dataSet =
    buildDataSet conn
      "SELECT Name, Number, Weight from Elements";;
val dataSet : DataSet
```

The following extracts the `DataTable` of results from this `DataSet` and iterates over the rows printing the results:

```
> let table = dataSet.Tables.Item 0;;
val table : DataTable
> for row in table.Rows do
    printf "%A\n" (row.Item "Name",
                   row.Item "Number",
                   row.Item "Weight");;
("Hydrogen", 1, 1.008)
("Helium", 2, 4.003)
val it : unit = ()
```

Note how the value of the field with the string name `field` in the row `row` is obtained using *row* . `Item` *field*.

In addition to programmatically enumerating over the results of a query, the `DataSet` can be used to inject data into the database programmatically as well. The following creates a new row in the table and populates it with the data for Lithium:

```
> let row = table.NewRow();;
val row : DataRow
> row.Item "Name" <- "Lithium";;
val it : unit = ()
> row.Item "Number" <- 3;;
val it : unit = ()
```

```
> row.Item "Weight" <- 6.941;;
val it : unit = ()
> table.Rows.Add row;;
val it : unit = ()
```

This change can be uploaded to the database using the `Update` member of the `SqlDataAdapter`:

```
> dataAdapter.Update dataSet;;
val it : int = 1
```

The return value of 1 indicates that a single row was altered.

Querying the database again shows that it does indeed now contain three rows:

```
> query();;
val it : seq<string * int * float>
= seq [("Hydrogen", 1, 1.008); ("Helium", 2, 4.003);
       ("Lithium", 3, 6.941)]
```

Databases are much more useful when they are filled by a program.

10.3.5 Filling the database from a data structure

Now we're ready to inject data about all of the chemical elements into the database programmatically. We begin by deleting the three existing rows to avoid conflicts:

```
> execNonQuery conn "DELETE FROM Elements";;
val it : unit = ()
```

The following loop adds the data for each element (assuming the existence of a data structure `elements` that is a sequence of records with the appropriate fields) to the table and then uploads the result:

```
> for element in elements do
    let row = table.NewRow()
    row.Item "Name" <- element.name
    row.Item "Number" <- element.number
    row.Item "Weight" <- element.weight
    table.Rows.Add row;;
val it : unit = ()
> dataAdapter.Update dataSet |> ignore;;
val it : unit = ()
```

The database now contains information about the chemical elements from this data structure.

10.3.6 Visualizing the result

As an industrial-strength platform, .NET naturally makes it as easy as possible to visualize the contents of a database table.

We begin by creating a blank Windows Form:

```
> open System.Windows.Forms;;
> let form = new Form(Text="Elements", Visible=true);;
val form : Form
```

As usual, forcing the form to stay on top is useful when developing in an interactive session:

```
> form.TopMost <- true;;
val it : unit = ()
```

The `DataGrid` class provides a Windows Forms control that can be bound to a database table in order to visualize it interactively:

```
> let grid = new DataGrid(DataSource=table);;
val grid : DataGrid
> grid.Dock <- DockStyle.Fill;;
val it : unit = ()
```

Note that the grid was bound to our database table and its dock style was set to fill the whole form when it is added:

```
> form.Controls.Add grid;;
val it : unit = ()
```

This tiny amount of code produces the interactive GUI application illustrated in figure 10.2.

10.3.7   Cleaning up

One of the key advantages of using a database is persistence: the contents of the database will still be here the next time we restart F# or Visual Studio or even the machine itself. However, we inevitably want to delete our rows, tables and databases. This line deletes the `Elements` table:

```
> execNonQuery conn "DROP TABLE Elements";;
val it : unit = ()
```

And this line deletes the database itself:

```
> execNonQuery conn "DROP DATABASE chemicalelements";;
val it : unit = ()
```

This chapter has shown how web services can be consumed easily in F# programs by reusing the capabilities provided for C#, and how databases can be created and used from F# programs with minimal effort.

Web applications and databases are the bread and butter of the .NET platform and a great many C# and Visual Basic programs already use this technology. However, F# is the first modern functional programming language to provide professional-quality web and database functionality and, consequently, is opening new avenues

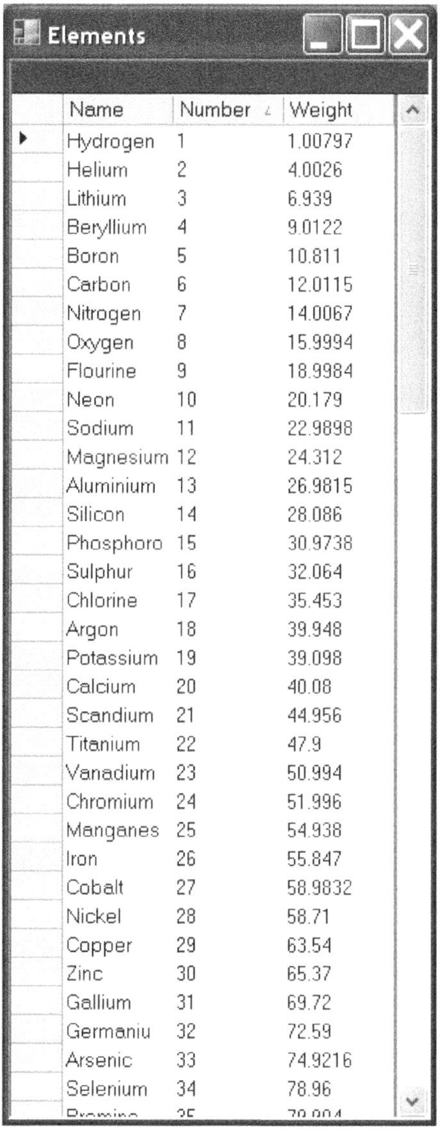

	Name	Number ∠	Weight	
▶	Hydrogen	1	1.00797	
	Helium	2	4.0026	
	Lithium	3	6.939	
	Beryllium	4	9.0122	
	Boron	5	10.811	
	Carbon	6	12.0115	
	Nitrogen	7	14.0067	
	Oxygen	8	15.9994	
	Flourine	9	18.9984	
	Neon	10	20.179	
	Sodium	11	22.9898	
	Magnesium	12	24.312	
	Aluminium	13	26.9815	
	Silicon	14	28.086	
	Phosphoro	15	30.9738	
	Sulphur	16	32.064	
	Chlorine	17	35.453	
	Argon	18	39.948	
	Potassium	19	39.098	
	Calcium	20	40.08	
	Scandium	21	44.956	
	Titanium	22	47.9	
	Vanadium	23	50.994	
	Chromium	24	51.996	
	Manganes	25	54.938	
	Iron	26	55.847	
	Cobalt	27	58.9832	
	Nickel	28	58.71	
	Copper	29	63.54	
	Zinc	30	65.37	
	Gallium	31	69.72	
	Germaniu	32	72.59	
	Arsenic	33	74.9216	
	Selenium	34	78.96	
	Bromine	35	79.904	

Figure 10.2 Visualizing the contents of a SQL Server database using the Windows Forms `DataGrid` control.

for combining these techniques. Functional programming will doubtless play an increasingly important role in web and database programming just as it is changing the way we think about other areas of programming.

CHAPTER 11

INTEROPERABILITY

Modern scientific computing often requires many separate components to interact. These components typically use different styles, are written in different languages and sometimes even run on separate platforms or architectures. The F# programming language provides a unique combination of expressive power and easy interoperability. This chapter is devoted to explaining just how easily F# allows programs to interact with other systems and even other platforms across the internet.

Due to the wide variety of different software used by scientists, a breakdown of the different approaches to interoperation is useful as the exact use of each and every package is beyond the scope of this book. This chapter illustrates how COM and .NET applications can be interoperated with using three of the most important applications: Microsoft Excel, The Mathwork's MATLAB and Wolfram Research's Mathematica.

11.1 EXCEL

The .NET platform is based upon the Common Language Runtime (**CLR**). The primary benefit of this design is the astonishing ease with which different .NET programs can interoperate. The simultaneous and interactive use of Microsoft's

Excel spreadsheet application and the F# programming language is no exception. The only non-trivial aspect of interoperating with other .NET applications from .NET languages like F# is the use of dynamically typed interfaces in many cases, including the interfaces to Microsoft Office applications such as Excel.

Microsoft Office is an almost ubiquitous piece of software, found on most of the world's desktop computers. Among the components of Office, Microsoft Excel is probably the most valuable for a scientist. Spreadsheets are deceptively powerful and Excel's unique graphical user interface facilitates the construction of complicated computations in a purely functional form of programming. However, when computations become too time consuming or complicated, or require the use of more advanced programming constructs and data structures, solutions written in Excel can be productively migrated to more suitable tools such as the F# programming language. The pairing of F# and Excel rivals the capabilities of many expensive technical computing environments for practical data acquisition and analysis and the ability to interoperate between F# and Excel is pivotal.

This section explains just how easily F# and Excel can interoperate by injecting data from F# programs directly into running Excel spreadsheets and reading results back.

11.1.1   Referencing the Excel interface

The interface required to use Excel is in the "Excel.dll" file. This is already on the search path for DLLs, so it can be loaded in a single line:

```
> #r "Excel.dll";;
```

In Office 2003, the interface is held in the `Excel` namespace:

```
> open Excel;;
```

Later version of Office use the `Microsoft.Office.Interop.Excel` namespace.

Before Excel can be used from other .NET languages such as F#, a handle to a running instance of Excel must be obtained. This is most easily done by creating a new instance of Excel and keeping the handle to it:

```
> let app = new ApplicationClass(Visible = true);;
val app : ApplicationClass
```

In order to manipulate a spreadsheet, a *workbook* must be either loaded from file or created afresh.

11.1.2   Loading an existing spreadsheet

A spreadsheet related to a given F# project is typically stored in the same directory as the source code of the F# project:

```
> let file = __SOURCE_DIRECTORY__ + @"\Example.xls";;
```

```
val file : string
```

A spreadsheet file ("Example.xls" in this case) can be loaded into the running instance of Excel using[21]:

```
> let workbook =
    let u = System.Reflection.Missing.Value
    app.Workbooks.Open(file, u, u, u, u, u, u,
                        u, u, u, u, u, u, u, u);;
val workbook : Workbook
```

This makes it easy to use F# and Excel both interactively and concurrently by storing the spreadsheets and programs together in the same directory.

11.1.3   Creating a new spreadsheet

For simple examples or temporary uses of Excel from F#, the creation of a new spreadsheet inside Excel can be automated by F# code:

```
> let workbook =
    app.Workbooks.Add(XlWBATemplate.xlWBATWorksheet);;
val workbook : Workbook
```

Once a spreadsheet has been loaded or created in Excel, it can be manipulated from F#.

11.1.4   Referring to a worksheet

In order to edit the cells in a particular worksheet it is necessary to obtain a reference to the worksheet itself. This is done by extracting the sequence of worksheets from the workbook and then choosing the appropriate one, such as the first one:

```
> let worksheet =
    workbook.Worksheets.[box 1] :?> Worksheet;;
val worksheet : Worksheet
```

Note the use of box to convert a statically-typed F# value 1 into a dynamically-typed .NET class obj and :?> to perform a run-time type tested downcast from the obj class to the Worksheet class.

This worksheet object provides member functions that allow a wide variety of properties to be set and actions to be invoked. The remainder of this section is devoted to using the Cells member of this object to get and set the values of spreadsheet cells, the simplest way for F# programs to interact with Excel spreadsheets.

---

[21]The F# designers have indicated that it will be possible to omit the "u" arguments in a later release of the language.

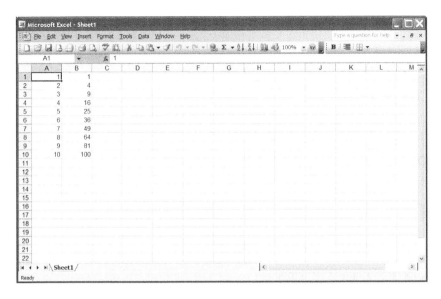

Figure 11.1 An Excel spreadsheet with cell values generated from a F# interactive session.

11.1.5 Writing cell values into a worksheet

The expression `worksheet.Cells.Item(i, j)` gives the contents of the cell in the ith row and j^{th} column. For example, $(i, j) = (3, 5)$ corresponds to the cell E3 in the spreadsheet. Note that the rows and columns are indexed starting from one rather than zero.

The following curried `set` function for setting a cell in a worksheet is very useful:

```
> let set (j : int) (i : int) v =
    worksheet.Cells.Item(i, j) <- box v;;
val set : int -> int -> 'a -> unit
```

Note the use of `box` to a given dynamically-typed value and the use of explicit type annotations, restricting `i` and `j` to be of the type `int`, to improve static type checking of subsequent F# code.

The argument order of this `set` function was chosen such that the column index can be partially applied first, because accessing a column (rather than a row) is the most common mode of use for functions that access spreadsheet cells.

```
> for i in 1 .. 10 do
    set 1 i i
    set 2 i (i*i);;
```

The resulting two-column spreadsheet is illustrated in figure 11.1.

When spreadsheet cells are set in this way, Excel automatically updates everything in the spreadsheet as required and there is no need to call an explicit update function.

11.1.6   Reading cell values from a worksheet

The writing of cell values to a worksheet was simplified by the interface automatically casting reasonably-typed values into the appropriate form for Excel. Reading values from a worksheet is slightly more complicated because the values are run-time typed and, before an F# program can perform any computations upon them, they must be cast to an appropriate static type.

The following curried `get` function reads a worksheet cell:

```
> let get (j : int) (i : int) =
    (worksheet.Cells.Item(i, j) :?> Range).Value2;;
val get : int -> int -> obj
```

Note that the return type of this function is `obj`. This is the universal class in .NET, the class that all other classes derive from. Consequently, the fact that a value is of the `obj` class conveys almost no information about the value. Before values such as the result of calling this `get` function can be used in F# programs, they must be cast into a static type such as `int`, `float` or `string`.

For example, this get function may be used to read the contents of the cell B5:

```
> get 2 5;;
val it : obj = 25.0
```

The result is correctly displayed as a `float` by the F# interactive session but the type must be ossified before the value can be used in F# programs.

The following `get_float` function uses the previously-defined `get` function and casts the result into a floating point number:

```
> let get_float j i =
    get j i :?> float;;
val get_float : int -> int -> float
```

For example, the value of the spreadsheet cell B5 is now correctly converted into a `float`:

```
> get_float 2 5;;
val it : float = 25.0
```

A `NullReferenceException` is thrown when attempting to get the `float` value of an empty cell because empty cells are represented by the `null` value:

```
> get_float 3 1;;
System.NullReferenceException: Object reference not set
to an instance of an object.
```

An `InvalidCastException` is thrown when attempting to get the `float` value of a cell that contains a value of another type because the run-time type conversion in `:?>` fails. For example, when trying to extract a `string` using the `get_float` function:

```
> set_float 3 1 "foo";;
```

```
val it : unit = ()
> get_float 3 1;;
System.InvalidCastException: Specified cast is not
valid.
```

The ability to read and write cells in Excel spreadsheets from F# programs is the foundation of practical interoperation between these two systems. The practical applications facilitated by this ease of interoperability are far more than the capabilities of either Excel or F# alone.

This section has not only described interoperation with Excel in detail but has also paved the way for interoperating with other .NET applications, including Microsoft Word. All .NET applications provide similar interfaces and are just as easy to use once the implications of dynamically-typed interfaces and run-time type casting are understood.

11.2   MATLAB

The Windows platform is home to a wide variety of scientific software. Many of these applications have not yet been updated for the .NET era and provide slightly more old-fashioned interfaces. By far the most common such interface is the Component Object Model (**COM**). This is in many ways a predecessor to .NET. This section describes the use of COM interfaces from .NET programming languages like F#, with a particular focus on one of the most popular applications among scientists and engineers: MATLAB.

MATLAB is a high-level language and interactive environment designed to enable its users to perform computationally intensive tasks more easily than with traditional programming languages such as C, C++, and Fortran.

The easiest way to use MATLAB from F# is via its Component Object Model (**COM**) interface. The Type Library Importer (tlbimp) command-line tool converts the type definitions found within a COM type library into equivalent definitions in a common language runtime assembly suitable for .NET. The output is a binary file (an assembly) that contains runtime metadata for the types defined within the original type library. You can examine this file with tools such as ildasm. The tlbimp command-line tool is part of the Microsoft .NET SDK.

11.2.1   Creating a .NET interface from a COM interface

From a DOS prompt, the following command enters the directory of the MATLAB "mlapp.tlb" file and uses Microsoft's tlbimp tool to create a .NET DLL implementing the MATLAB interface:

```
C:\> cd "C:\Program Files\MATLAB\R2007a\bin\win32\"
C:\> "C:\Program Files\Microsoft.NET\SDK\v2.0\Bin\
tlbimp.exe" mlapp.tlb
```

The same procedure can be used to compile DLLs providing .NET interfaces to many COM libraries.

11.2.2   Using the interface

In F#, the directory containing the new "MLApp.dll" file can be added to the search path in order to load the DLL:

```
> #I "C:\Program Files\MATLAB\R2007a\bin\win32\";;
> #r "MLApp.dll";;
```

A fresh instance of MATLAB can be started and a handle to it obtained in F# using:

```
> let matlab = new MLApp.MLAppClass();;
val matlab : MLApp.MLAppClass
```

Once the .NET interface to MATLAB has been created, loaded and instantiated in F# the two systems are able to interoperate in a variety of ways. The simplest form of interoperability is simply invoking MATLAB commands remotely from F#. This can be useful for creating diagrams but the interface also allows arbitrary data to be transferred between the two systems by reading and writing the values of MATLAB variables.

11.2.3   Remote execution of MATLAB commands

The new MATLAB instance can be controlled by invoking commands from F# by passing a string to the Execute member function of the matlab class. For example, the following creates a simple 2D function plot:

```
> matlab.Execute "x = 0:0.01:2*pi; plot (x, sin(x))";;
val it : string = ""
```

The result is illustrated in figure 11.2.

11.2.4   Reading and writing MATLAB variables

In addition to being able to interactively invoke commands in MATLAB from F#, the ability to get and set the values of MATLAB variables from F# programs is also useful. The following functions get and set MATLAB variables of arbitrary types:

```
> let get name =
    matlab.GetVariable(name, "base");;
val get : string -> obj
> let set name value =
    matlab.PutWorkspaceData(name, "base", value);;
val set : string -> 'a -> unit
```

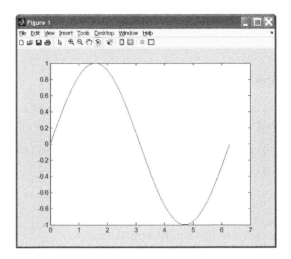

Figure 11.2   A plot of $sin(x)$ created in MATLAB by an F# program.

As this interface is dynamically typed and the MATLAB language supports only a small number of types, it is useful to provide some get and set functions with specific types, most notably `float` for numbers and `float [,]` for vectors and matrices:

```
> let get_float name =
    get name :?> float;;
val get_float : string -> float
> let get_vecmat name =
    get name :?> float [,];;
val get_vecmat : string -> float [,]
```

Numbers can be set directly using the `set` function but matrices of the .NET type `float [,]` require explicit construction because there are no literals for this type in the F# language. In this context, a function to convert an arbitrary sequence of sequences into a 2D .NET array is useful:

```
> let array2_of a =
    let a = Array.of_seq (Seq.map Array.of_seq a)
    Array2.init (Array.length a) (Array.length a.[0])
      (fun i j -> a.[i].[j]);;
val array2_of : #seq<'b> -> 'c [,] when 'b :> seq<'c>
```

For example, setting the variable v to the row vector $(1, 2, 3)$ in MATLAB from F#:

```
> set "v" (array2_of [ [ 1.0; 2.0; 3.0 ] ]);;
> get_vecmat "v";;
val it : float [,] = [|[|1.0; 2.0; 3.0|]|]
```

The ability to invoke arbitrary commands in MATLAB as well as read and write variable names allows new F# programs to interoperate seamlessly with existing MATLAB programs and provides a new dimensionality to the function of these systems.

11.3 MATHEMATICA

This technical computing environment from Wolfram Research is particularly useful for symbolic computation as it is built around a fast term rewriter with a huge standard library of rules for solving mathematical problems symbolically. Interoperability between Mathematica and F# is particularly useful in the context of numerical programs written in F# that involve the computation of symbolic expressions generated by Mathematica. This is most easily achieved using the excellent .NET-link interface to Mathematica that allows .NET programs to interoperate with Mathematica.

11.3.1 Using .NET-link

The .NET-link interface to Mathematica 6 may be loaded directly as a DLL:

```
> #light;;
> #I @"C:\Program Files\Wolfram Research\Mathematica\
6.0\SystemFiles\Links\NETLink";;
> #r "Wolfram.NETLink.dll";;
```

The interface is provided in the following namespace:

```
> open Wolfram.NETLink;;
```

The definitions in this namespace allow Mathematica kernels to be spawned and interoperated with to perform symbolic computations and extract the results in symbolic form. This is particularly useful in the context of high-performance evaluation of symbolic expressions because Mathematica excels at manipulating mathematical expressions and F# excels at high-performance evaluation of symbolic expressions.

The F# definitions pertaining to these symbolic expressions use definitions from the `Math` namespace:

```
> open Math;;
```

A variant type can be used to represent a symbolic expression:

```
> type expr =
    | Integer of int
    | Symbol of string
    | ArcTan of expr
    | Log of expr
    | Tan of expr
    | Plus of expr * expr
```

```
    | Power of expr * expr
    | Times of expr * expr
    | Rational of expr * expr;;
```

The following function converts a string and sequence of expressions into a function expression, handling the associativities of the operators when building expression trees:

```
> let rec func h t =
    match h, t with
    | "ArcTan", [f] -> ArcTan f
    | "Log", [f] -> Log f
    | "Tan", [f] -> Tan f
    | "Plus", [] -> Integer 0
    | ("Times" | "Power"), [] -> Integer 1
    | ("Plus" | "Times" | "Power"), [f] -> f
    | "Plus", [f; g] -> Plus(f, g)
    | "Times", [f; g] -> Times(f, g)
    | "Power", [f; g] -> Power(f, g)
    | ("Plus" | "Times" as h), f::fs ->
        func h [f; func h fs]
    | "Power", fs ->
        List.fold1_right (fun f g -> func h [f; g]) fs
    | "Rational", [p; q] -> Rational(p, q)
    | h, _ -> invalid_arg("func " + h);;
val func : string -> expr list -> expr
```

The following function tries to read a Mathematica expression, using the func function to convert the string representation of function applications used by Mathematica into the appropriate constructor of the expr type:

```
> let rec read (ml : IKernelLink) =
    match ml.GetExpressionType() with
    | ExpressionType.Function ->
        let args = ref 0
        let f = ml.GetFunction(args)
        func f [for i in 1 .. !args -> read ml]
    | ExpressionType.Integer -> Integer(ml.GetInteger())
    | ExpressionType.Symbol -> Symbol(ml.GetSymbol())
    | ExpressionType.Real -> invalid_arg "read real"
    | ExpressionType.String -> invalid_arg "read string"
    | ExpressionType.Boolean -> invalid_arg "read bool"
    | _ -> invalid_arg "read";;
val read : Wolfram.NETLink.IKernelLink -> expr
```

The following mma function spawns a new Mathematica kernel and uses the link to evaluate the given expression and read back the symbolic result:

```
> let mma (expr : string) =
    let ml = MathLinkFactory.CreateKernelLink()
    try
      ml.WaitAndDiscardAnswer()
      ml.Evaluate(expr)
      ml.GetFunction(ref 0) |> ignore
      read ml
    finally
      ml.Close();;
val mma : Wolfram.NETLink.IKernelLink -> string -> expr
```

This function is careful to discard the handshake message inserted after the link is made and the `ReturnPacket` function call that wraps a valid response.

Although spawned kernels can be reused, the performance overhead of spawning a custom Mathematica kernel is insignificant for our example and this approach evades problems caused by incorrectly parsed results introducing synchronization problems with the kernel. The `getExpr` member provides a higher-level interface that replaces our use of `getFunction` and friends. However, at the time of writing the `Expr` representation provided by Mathematica is designed to allow symbolic expressions to be composed and injected into Mathematica rather than extracted from Mathematica, which is the application of our example. Thus, we use the lower-level interface.

For example, adding two symbols in Mathematica and reading the symbolic result back gives a value of the variant type `expr` in F#:

```
> mma "a+b";;
val it : expr = Plus (Symbol "a",Symbol "b")
```

As we have already seen, symbolic expressions represented in this form can be manipulated very simply and efficiently by F# programs.

11.3.2 Example

Consider the result of the indefinite integral:

$$f(x) = \int \sqrt{\tan x}\,dx$$

Mathematica is able to take this integral symbolically but is slow to evaluate the complicated resulting expression. Moving the symbolic result into F# allows it to be evaluated much more efficiently:

```
> let f = mma "Integrate[Sqrt[Tan[x]], x]";;
val f : expr
```

The resulting expression is quite complicated:

```
> f;;
val it : expr
= Times
```

```
(Rational (Integer 1,Integer 2),
 Times
   (Power (Integer 2,Rational (Integer -1,Integer 2)),
    Plus
      (Times
        (Integer -2,
```
. . .

Mathematica is able to evaluate this expression for complex values of x 360,000 times in 26 seconds. A simple F# function is able to evaluate this expression much more efficiently.

Evaluating this expression in F# requires several functions over complex numbers that are provided by the F# standard library and some functions that are not:

```
> open Math.Complex;;
```

The pow function raises one complex number to the power of another and may be written in terms of exp and log:

$$z_1^{z_2} = e^{z_2 \ln z_1}$$

```
> let pow z1 z2 =
    exp(z2 * log z1);;
val pow : Complex -> Complex -> Complex
```

The arc tangent of $\frac{y}{x}$ may be expressed as:

$$\tan^{-1}\left(\frac{y}{x}\right) = -i\frac{\ln(x + iy)}{\sqrt{x^2 + y^2}}$$

```
> let atan2 x y =
    -onei * log((x + onei * y) / sqrt(x * x + y * y));;
val atan2 : Complex -> Complex -> Complex

> let atan z =
    atan2 one z;;
val atan : Complex -> Complex
```

A value of the expr type may be evaluated in the context of a mapping from symbol names to complex values implemented by subst using the following eval function:

```
> let rec eval subst f =
    match f with
    | ArcTan f -> atan(eval subst f)
    | Log f -> log(eval subst f)
    | Plus(f, g) -> eval subst f + eval subst g
    | Power(f, g) -> pow (eval subst f) (eval subst g)
    | Times(f, g) -> eval subst f * eval subst g
```

```
    | Rational(p, q) -> eval subst p / eval subst q
    | Tan f -> tan(eval subst f)
    | Integer n -> complex (float n) 0.0
    | Symbol s -> subst s;;
val eval : (string -> Complex) -> expr -> Complex
```

Evaluating the example expression f in the context $x = 0.1 + 0.3i$ gives $f(x) = -0.036 + 1.22i$ as expected:

```
> eval (function
            | "x" -> complex 0.1 0.3
            | _ -> zero) f;;
val it : Complex = -0.03561753878r+1.22309153i
```

The following function computes the same tabulation of complex values of this symbolic expression that Mathematica took 26s to evaluate:

```
> let gen f =
    [|for x in -3.0 .. 0.01 .. 3.0 ->
        [|for y in -3.0 .. 0.01 .. 3.0 ->
            eval (function
                    | "x" -> complex 0.1 0.1
                    | _ -> zero) f |] |]
val gen : expr -> Math.complex array array
```

Despite its simplicity, the F# evaluator is able to evaluate the same result $3.4\times$ faster than Mathematica:

```
> let data = time gen f;;
Took 7595ms
```

Given that Mathematica is specifically designed for manipulating symbolic expressions, it might be surprising that such a considerable performance improvement can be obtained simply by writing what is little more than part of Mathematica's own expression evaluator in F#. The single most important reason for this speed boost is the specialization of the F# code compared to Mathematica's own general-purpose term rewriter. The representation of a symbolic expression as a value of the expr type in this F# code only handles nine different kinds of expression whereas Mathematica handles an infinite variety, including arrays of expressions.

Moreover, the F# programming language also excels at compiler writing and the JIT-compilation capabilities of the .NET platform make it ideally suited to the construction of custom evaluators that are compiled down to native code before being executed. This approach is typically orders of magnitude faster than evaluation in a standalone generic term rewriting system like Mathematica.

The marriage of Mathematica's awesome symbolic capabilities with the performance and interoperability of F# makes a formidable team for any applications where complicated symbolic calculations are evaluated in a computationally intensive way.

CHAPTER 12

COMPLETE EXAMPLES

This chapter details several complete programs that demonstrate some of the most importants forms of scientific computing whilst also leveraging the elegance of the F# language and the power of the .NET platform.

12.1   FAST FOURIER TRANSFORM

The program developed in this section combines a core concept in scientific computing with a core concept in computer science:

- Spectral analysis: computing the Fourier transform.

- Divide and conquer algorithms.

The Fourier transform is one of the most essential tools in scientific computing, with applications in all major branches of science and engineering as well as computer science and even mathematics. This section describes the development of an efficient implementation of the Fourier transform known as the Fast Fourier Transform (**FFT**).

12.1.1 Discrete Fourier transform

In its simplest form, the Fourier transform $\hat{f}_k$ of a signal f_n may be written:

$$\hat{f}_k = \sum_{n=0}^{N-1} f_n e^{-\frac{2\pi i}{N} nk}$$

This direct summation algorithm is referred to as the Discrete Fourier Transform (**DFT**) and is composed of $O(N^2)$ operations, i.e. this algorithm has quadratic asymptotic time complexity.

This naive algorithm may be implemented as a simple F# function:

```
> #light;;

> open System;;

> open Math;;

> let dft ts =
    let N = Array.length ts
    [|for k in 0 .. Array.length ts - 1 ->
       let mutable z = Complex.zero
       for n=0 to N-1 do
         let w =
           2.0 * Math.PI / float N * float n * float k
         z <- z + ts.[n] * complex (cos w) (sin w)
       z|];;
val dft : Complex array -> complex array
```

For example, the dft function correctly computes the Fourier series of a sine wave:

```
> [| for i in 0 .. 15 ->
      complex (float i / 8.0 * Math.PI |> sin) 0.0|]
  |> dft;;
val it : complex array =
  [|0; 8i; 0; 0; 0; 0; 0; 0; 0; 0; 0; 0; 0; 0; 0; -8i|]
```

This output assumes the use of the pretty printer for complex numbers given in section 9.11.3.

However, this implementation is very slow, taking 5.8s to compute a simple 2^{12}-point DFT:

```
> [| for n in 1 .. 4096 ->
      complex (float n) 0.0 |]
  |> time (ignore << dft);;
Took 5792ms
```

As we shall see, the performance of this implementation can be greatly improved upon.

12.1.2 Danielson-Lanczos algorithm

As always, algorithmic optimizations are the best place to search for performance improvements. In this case, the naive $O(N^2)$ algorithm may be replaced by an $O(N \log N)$ divide-and-conquer algorithm when N is an integral power of two. This algorithm, often referred to as the radix-2 Fast Fourier Transform (**FFT**) has been the foundation of numerical Fourier transforms for at least 200 years. The FFT algorithm splits an n-point FFT into two $\frac{1}{2}n$-point FFTs composed of the even- and odd-indexed elements, respectively, as follows:

$$\hat{f}_k = \sum_{n=0}^{N-1} f_n e^{-\frac{2\pi i}{N}nk}$$

$$= \sum_{n=0}^{\frac{1}{2}N-1} f_{2n} e^{-\frac{2\pi i}{N}2nk} + e^{-\frac{2\pi i}{N}k} \sum_{n=0}^{\frac{1}{2}N-1} f_{2n+1} e^{-\frac{2\pi i}{N}(2n+1)k}$$

The prefactors $e^{-\frac{2\pi i}{N}k}$ are complex roots of unity and are called *twiddle factors*.

Using conventional rearrangements, this algorithm may be decomposed into an initial shuffle of the input followed by a sequence of traversals over the input data.

The sort and subsequent rewrites of the array are most easily written in terms of swaps and in-place rewrites, both of which may be elegantly added to the built-in `Array` module:

```
> module Array =
    let swap (a : 'a array) i j =
      let t = a.[i]
      a.[i] <- a.[j]
      a.[j] <- t

    let apply f a =
      for i=0 to Array.length a-1 do
        a.[i] <- f a.[i];;
module Array : begin
  val swap : 'a array -> int -> int -> unit
  val apply : ('a -> 'a) -> 'a array -> unit
end
```

An initial sort can reorder the elements of the input array in lexicographic order by the bits of their indices. This may be implemented as a simple F# function that uses bit-wise integer operations:

```
> let bitrev a =
    let n = Array.length a
    let mutable j = 0
    for i=0 to n-2 do
      if i<j then Array.swap a i j
      let rec aux m j =
```

```
        let m = m/2
        let j = j ^^^ m
        if j &&& m = 0 then aux m j else j
    j <- aux n j;;
val bitrev : 'a array -> unit
```

The inner loop of the algorithm operates on a `Complex` array and is easily written in F# using its overloaded operators to handle complex arithmetic:

```
> let fft_aux (a : Complex array) n j sign m =
    let w =
      let t = Math.PI * float (sign * m) / float j
      complex (cos t) (sin t)
    let mutable i = m
    while i < n do
      let ai = a.[i]
      let t = w * a.[i + j]
      a.[i] <- ai + t
      a.[i + j] <- ai - t
      i <- i + 2*j;;
val fft_aux :
  Complex array -> int -> int -> int -> int -> unit
```

The following `fft_pow2` function uses the `bitrev` and `loop` functions to compute the FFT of vectors with lengths that are integral powers of two:

```
> let fft_pow2 sign a =
    let n = Array.length a
    bitrev a
    let mutable j = 1
    while j < n do
      for m = 0 to j - 1 do
        fft_aux a n j sign m
      j <- 2 * j;;
val fft_pow2 : int -> Complex array -> unit
```

The previous example of a 2^{12}-point transform that took 5.8s with the naive $O(N^2)$ algorithm now only takes 0.009s:

```
> [| for n in 1 .. 4096 ->
       complex (float n) 0.0 |]
  |> time (ignore << fft_pow2 1);;
Took 9ms
```

In fact, this divide-and-conquer algorithm is so much faster that it can solve problems that were completely infeasible with the previous implementation:

```
> [| for n in 1 .. 1048576 ->
       complex (float n) 0.0 |]
```

```
|> time (ignore << fft_pow2 1);;
Took 4781ms
```

However, this implementation can only be applied when N is an integral power of two.

12.1.3 Bluestein's convolution algorithm

The Fourier transform of an arbitrary-length signal may be computed by rephrasing the transform as a convolution and zero padding the convolution up to an integral power of two length. Note that zero padding a convolution is exact.

The DFT may be expressed as a convolution:

$$
\hat{f}_k = \sum_{n=0}^{N-1} f_n e^{-\frac{2\pi i}{N} nk}
$$

$$
= e^{-\frac{\pi i}{N} k^2} \sum_{n=0}^{N-1} \left(f_n e^{-\frac{\pi i}{N} n^2} \right) e^{-\frac{\pi i}{N} (k-n)^2}
$$

$$
= b_k^* \sum_{n=0}^{N-1} a_n b_{k-n}
$$

where:

$$
a = f_n e^{-\frac{\pi i}{N} n^2}
$$

$$
b = e^{\frac{\pi i}{N} n^2}
$$

The factors $e^{\frac{\pi i}{N} n^2}$ can be compiled into an array:

```
> let bluestein_sequence n =
    let s = Math.PI / float n
    [|for k in 0 .. n-1 ->
        let t = s * float(k * k)
        complex (cos t) (sin t)|];;
val bluestein_sequence : int -> complex array
```

The following function gives the next power of two above the given number:

```
> let rec next_pow2 = function
    | 0 -> 1
    | n -> 2 * next_pow2(n / 2);;
val next_pow2 : int -> int
```

Padding is accomplished using the following function:

```
> let pad n nb (w : complex array) = function
    | i when i < n -> w.[i]
    | i when i > nb - n -> w.[nb-i]
```

```
    | _ -> Complex.zero
val pad : int -> int -> complex array -> int -> complex
```

These can be combined to compute the Fourier transform of an arbitrary length signal in-place by zero padding its convolution up to an integral power of two and using the `fft_pow2` function:

```
> let bluestein a =
    let n = Array.length a
    let nb = pow2_atleast(2*n - 1)
    let w = bluestein_sequence n
    let y = pad n nb w |> Array.init nb
    fft_pow2 (-1) y
    let b = Array.create nb Complex.zero
    for i=0 to n-1 do
      b.[i] <- Complex.conjugate w.[i] * a.[i]
    fft_pow2 (-1) b
    let b = Array.map2 ( * ) b y
    fft_pow2 1 b
    let nbinv = complex (1.0 / float nb) 0.0
    for i=0 to n-1 do
      a.[i] <- nbinv * Complex.conjugate w.[i] * b.[i];;
val bluestein : Complex array -> unit
```

For the forward transform, the real and imaginary parts are exchanged using the following `swapri` function:

```
> let swapri a =
    Array.apply (fun z -> complex z.i z.r) a;;
val swapri : complex array -> unit
```

The following `is_pow2` function tests if a number is an integral power of two using the fact that only such numbers share no bits with their predecessor:

```
> let is_pow2 n =
    n &&& n-1 = 0;;
val is_pow2 : int -> bool
```

For example, filtering out the powers of two from a sequence of consecutive integers using the `is_pow2` function:

```
> List.filter is_pow2 [1 .. 1000];;
val it : int list =
  [1; 2; 4; 8; 16; 32; 64; 128; 256; 512]
```

These routines may be combined to compute an arbitrary FFT:

```
> let fft sign a =
    let n = Array.length a
    if is_pow2 n then fft_pow2 sign a else
```

```
    if sign=1 then swapri a
    bluestein a
    if sign=1 then swapri a;;
val fft : int -> Complex array -> unit
```

This fft function may be applied to arbitrary-length complex arrays to compute the Fourier transform in $O(N \log N)$ time using the given sign to determine the direction of the transform. Consequently, it is productive to wrap this function in fourier and ifourier functions that compute forward and reverse transforms using the appropriate signs:

```
> let fourier a =
    fft 1 a;
    a;;
val fourier : Complex array -> Complex array
> let ifourier a =
    fft (-1) a;
    a;;
val ifourier : Complex array -> Complex array
```

Note that these functions return the input array because this is often useful and not because they are out-of-place transforms.

12.1.4   Testing and performance

The new fourier function is most easily tested by comparing its input with that of the dft function on random input:

```
> let rand = new System.Random();;
val rand : System.Random
> let a =
    [|for i in 0 .. 14 ->
        complex
            (rand.NextDouble())
            (rand.NextDouble())|];;
val a : Complex array
> fourier (Array.copy a)
  |> Seq.map2 ( - ) (dft (Array.copy a))
  |> Seq.map Complex.magnitude;;
  |> Seq.fold max 0.0;;
val it : complex
= 1.754046975e-14
```

The maximum numerical error of 1.75×10^{-14} introduced by the more efficient fourier function is clearly very small.

This implementation remains very fast:

```
> [|for n in 1 .. 4096 ->
```

```
      complex (float n) 0.0|]
  |> time (ignore << fourier);;
Took 9ms
> [|for n in 1 .. 1048576 ->
      complex (float n) 0.0|]
  |> time (ignore << fourier);;
Took 4781ms
> [|for n in 1 .. 1048575 ->
      complex (float n) 0.0|]
  |> time (ignore << fourier);;
Took 30227ms
```

Despite having a smaller number of elements, the $2^{20} - 1$ length transform takes longer than the 2^{20} length transform because algorithm resorts to three separate 2^{n+1} length transforms instead.

As we have seen, the FFT is an excellent algorithm to introduce many of the most important concepts covered in the earlier chapters of this book. In particular, algorithmic optimization is a source of enormous performance improvements in this case.

12.2   SEMI-CIRCLE LAW

The program developed in this section combines two related subjects that are both core concepts in scientific computing:

- Linear algebra: eigenvalues.

- Random matrix theory.

The importance of linear algebra is widely known. Many naturally occurring phenomena can be modelled in terms of vectors and matrices. Most notable, perhaps, is the representation of quantum-mechanical operators as matrices, the eigenvalues of which are well known to have special importance [10]. Solving matrix problems can require various different forms of matrix manipulation, particularly forms of factorization. For example, one prevelant task is the computation of eigenvalues that we examine here. The eigenvalues of a Hamiltonian quantify the energy levels of the physical system.

Random matrix theory is a fascinating branch of mathematics dedicated to studying the properties of matrices with elements drawn from known probability distributions. Predictions about eigenvalue correlations and distributions are of great practical importance and random matrix theory has been responsible for several laws that turned out to be far more widely applicable than expected [2, 19].

In this section, we present an F# program that generates random matrices and uses a library to compute their eigenvalues before compiling the eigenvalue distribution and visualizing the results in Excel.

12.2.1 Eigenvalue computation

As discussed in chapter 9, the defacto-standard library for linear algebra on the .NET platform is the Extreme Optimization library from Numeric Edge. Consequently, we shall use this library to handle the generated matrix and compute its eigenvalues.

Before using the Extreme Optimization library, the DLL must be loaded:

```
> #light;;
> #I @"C:\Program Files\Extreme Optimization\
Numerical Libraries for .NET\bin";;
> #r "Extreme.Numerics.dll";;
```

Routines for handling symmetric matrices and computing their eigenvalues are in the following namespace:

```
> open Extreme.Mathematics.LinearAlgebra;;
```

The following `eigenvalues` function computes the eigenvalues of a symmetric $n \times n$ matrix with elements taken randomly from the set $\{-1, 1\}$:

```
> let eigenvalues n =
    let m = new SymmetricMatrix(n)
    let rand = new System.Random()
    for i=0 to n-1 do
      for j=i to n-1 do
        m.Item(i, j) <- float(2*rand.Next(2) - 1)
    m.GetEigenvalues();;
val eigenvalue : int -> GeneralVector
```

The `Item` property for the `SymmetricMatrix` class is overloaded and F# is unable to resolve the overload so the `a.[i] <- x` syntax cannot be used. This problem is circumvented by calling `a.Item(i) <- x` directly.

The eigenvalue solver in this library is so fast that a relatively large matrix can be used:

```
> let eigs = eigenvalues 1024;;
val eigs : Complex.ComplexGeneralVector
```

The following `get` function can be used to fetch an element from a map using the default value `0` for unspecified elements:

```
> let get (m : Map<'a, int>) i =
    try
      m.[i]
    with
    | Not_found ->
        0;;
val get : Map<'a,int> -> 'a -> int
```

This simple and asymptotically efficient implementation of a sparse container (e.g. a vector in this case) is adequate here because the collation of results is not performance critical. However, high-performance programs requiring faster sparse vector and matrix routines would benefit significantly from using an implementation such as that provided by the Extreme Optimization library.

The following `inc` function increments an element in a sparse map:

```
> let inc m i =
    Map.add i (1 + get m i) m;;
val inc : Map<'a,int> -> 'a -> Map<'a,int>
```

The following function accumulates the distribution of elements in a vector:

```
> let density_of (seq : GeneralVector) =
    let aux m x =
      x + 0.5 |> floor |> inc m
    Seq.fold aux Map.empty seq
    |> Map.to_array;;
val density_of : GeneralVector -> (float * int) array
```

The distribution of the eigenvalues `eigs` of the matrix is then given by:

```
> let density = density_of eigs;;
val density : (float * int) array
```

The results are most easily visualized by injecting them directly into a running Excel spreadsheet.

12.2.2 Injecting results into Excel

The simplest way to visualize the results generated by this program is using Excel. Fortunately, the Excel automation facilities provided by .NET make it very easy to inject results from an F# program directly into the cells of an Excel spreadsheet, as described in chapter 11.

The appropriate assembly must be loaded:

```
> #r "Excel.dll";;
```

The relevant functions are in the following namespaces:

```
> open Excel;;
```

```
> open System.Reflection;;
```

Injecting results into a spreadsheet requires an application class, workbook and worksheet:

```
> let app = new ApplicationClass(Visible = true);;
val app : ApplicationClass
```

A new workbook may be created with:

```
> let workbook =
    app.Workbooks.Add(XlWBATemplate.xlWBATWorksheet);;
```

```
val workbook : Workbook
```

The first worksheet in the spreadsheet is given by:

```
> let worksheet =
    (workbook.Worksheets.[box 1] :?> _Worksheet);;
val worksheet : _Worksheet
```

The following function allows a cell in the spreadsheet to be set:

```
> let set i j v =
    worksheet.Cells.Item(j, i) <- box v;;
val set : 'a -> 'b -> 'c -> unit
```

The following line iterates over the density of eigenvalues, filling in the spreadsheet cells with the value and density of eigenvalues:

```
> density
  |> Seq.iteri (fun j (a, b) ->
      set 1 (j+1) a
      set 2 (j+1) b);;
```

Excel can then be used to analyze and visualize the data.

12.2.3 Results

A famous result of random matrix theory is the semi-circle law of eigenvalue densities for random $n \times n$ matrices from the *Gaussian Orthogonal Ensemble* (**GOE**):

$$P(\lambda) = \begin{cases} \frac{2}{n\pi}\sqrt{n - \lambda^2} & -\sqrt{n} < \lambda < \sqrt{n} \\ 0 & \text{otherwise} \end{cases}$$

Although the derivation of the semi-circle law only applies to GOE matrices, the distributions of the eigenvalues found by this program (for $M_{ij} = \pm 1$) are also well approximated by the semi-circle law. The results of this program, illustrated in figure 12.1, demonstrate this.

12.3 FINDING N^{TH}-NEAREST NEIGHBORS

The program developed in this section combines four important algorithmic concepts that are commonly encountered in scientific computing:

- Set-theoretic operations: union, intersection and difference.

- Graph-theoretic operations: n^{th}-nearest neighbours.

- Implicit data structures: finite data represent an infinite graph.

- Dynamic programming: a form of divide and conquer that handles overlapping subproblems.

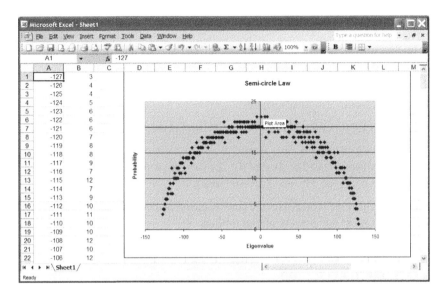

Figure 12.1   The approximately semi-circular eigenvalue density $P(\lambda)$ for a dense, random, square matrix $M_{ij} = \pm 1$ with $n = 1024$, showing the computed eigenvalue distribution.

- High-performance real-time interactive visualization that runs concurrently with the main computation.

The graph-theoretic problem of finding the nth-nearest neighbours allows useful topological information to be gathered from many forms of data produced by other scientific computations. For example, in the case of simulated atomic structures, where topological information can aid the interpretation of experimental results when trying to understand molecular structure. Such topological information can also be used indirectly, in the computation of interesting properties such as the decomposition of correlation functions over neighbour shells [5], and shortest-path ring statistics [7].

We shall describe our unconventional formulation of the problem of computing the nth-nearest neighbours of atoms in an atomic structure simulated under periodic boundary conditions before describing a program for solving this problem and presenting demonstrative results.

12.3.1 Formulation

The notion of the nth-nearest neighbours $\mathcal{N}_i^n$ of a vertex i in a graph is rigorously defined by a recurrence relation based upon the set of first nearest neighbours $\mathcal{N}_i^1 \equiv \mathcal{N}_i$ of any atom i:

$$\mathcal{N}_i^n = \begin{cases} \{i\} & n = 0 \\ \mathcal{N}_i & n = 1 \\ \left(\left(\bigcup_{j \in \mathcal{N}_i^{n-1}} \mathcal{N}_j\right) \setminus \mathcal{N}_i^{n-1}\right) \setminus \mathcal{N}_i^{n-2} & n \geq 2 \end{cases}$$

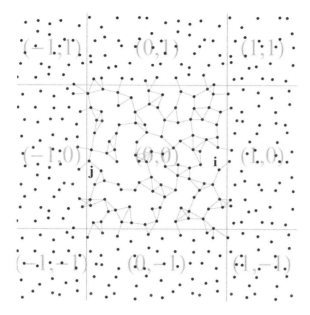

Figure 12.2   Conventionally, atoms are referenced only by their index $i \in \mathbb{I} = \{1 \dots N\}$ within the supercell. Consequently, atoms $i, j \in \mathbb{I}$ at opposite ends of the supercell are considered to be bonded.

As a recurrence relation, this computational task naturally lends itself to recursion. As this recurrence relation only makes use of the set-theoretic operations union and difference, the F# data structure manipulated by the recursive function is most naturally a set (described in section 3.4).

In order to develop a useful scientific program, we shall use an infinite graph to represent the topology of a d-dimensional crystal, i.e. a periodic tiling. Computer simulations of non-crystalline materials are typically formulated as a crystal with the largest possible unit cell, known as the *supercell*. Conventional approaches to the analysis of these structures reference atoms by their index $i \in \mathbb{I} = \{1 \dots N\}$ within the origin supercell. Edges in the graph representing bonded pairs of atoms in different cells are then handled by treating displacements modulo the supercell (illustrated in figure 12.2). However, this conventional approach is well-known to be flawed when applied to insufficiently large supercells [7, 8], requiring erroneous results to be identified and weeded out manually.

Instead, we shall choose to reference atoms by an index $i = (i_o, i_i)$ where $i_o \in \mathbb{Z}^d$ and $i_i \in \mathbb{I}$. This explicitly includes the offset i_o of the supercell as well as the index i_i within the supercell (illustrated in figure 12.3). Neighbouring vertices in the graph representing the topology are defined not only by the index of the neighbouring vertex but also by the supercell containing this neighbour (assuming the origin vertex to be in the origin supercell at $\{0\}^d$).

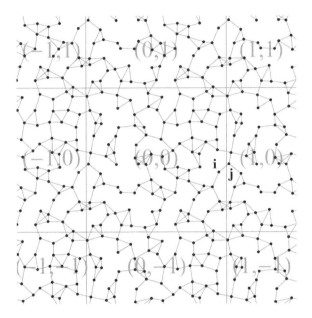

Figure 12.3   We use an unconventional representation that allows all atoms to be indexed by the supercell they are in as well as their index within the supercell. In this case, the pair of bonded atoms are referenced as $((0,0), i_i)$ and $((1,0), i_j)$, i.e. with i in the origin supercell $(0,0)$ and j in the supercell with offset $(1,0)$.

We shall now develop a complete program for computing the nth-nearest neighbours of a given vertex with index i from a list of lists $\mathbf{r}_i$ of the indices of the neighbours of each vertex.

12.3.2   Representing an atomic configuration

A common design pattern in programs that include lexers and parsers involves the parser and main program depending upon the definition of a core data structure. This pattern is most elegantly implemented by placing the definition of the data structure in a separate compilation unit that both the parser and the main program depend upon.

In this case, the data structure used to represent an atomic configuration must be shared between the parser and main program. Consequently, this data structure is defined in a separate compilation unit called "model.fs":

```
module Model
type coord = { index: int; offset: vector }
type vertex =
  { position: vector; neighbors: Set<coord> }
type t = { supercell: vector; vertices: vertex array }
```

The `coord` type represents a vertex in the infinite graph with the given integer index into the vertex array and the given supercell offset vector. The elements of the offset vector are integers but are stored as floating point numbers in this case simply because subsequent computations are made easier by the use of the uniform, built-in `vector` type.

The `vertex` type represents a vertex in the origin supercell of the graph and is quantified by the vector coordinate of the vertex and its set of neighbors.

The type `t` represents a complete model structure with the given supercell dimensions and array of vertices in the origin tile.

12.3.3   Parser

The format read by this program is in Mathematica syntax. Although the lexer and parser could handle their input in a generic way, performance can be significantly improved by specializing them to the particular structure of expression used to represent a model in order to avoid creating a large intermediate data structure, i.e. this is a deforesting optimization.

The parser begins by opening the `Model` module to simplify the use of its data structures:

```
%{
  open Model
%}
```

The types and values of tokens are then defined:

```
%token <float> REAL
```

```
%token COMMA OPEN CLOSE
```

The start point of the grammar and the type returned by its action are defined before the grammar itself is described:

```
%start system
%type <t> system
%%
```

In this case, the grammar is most easily broken down into vectors, neighbors, vertices and a whole system. A vector is taken to be a 3D vector written in Mathematica's List syntax:

```
vec3 :
| OPEN REAL COMMA REAL COMMA REAL CLOSE
    { vector [$2; $4; $6] }
;
```

A neighbor is composed of its position and index, again represented by nested Mathematica lists:

```
neighbor :
| OPEN vec3 COMMA REAL CLOSE
    { { offset = $2; index = int $4 - 1 } }
;
```

Note the use of the overloaded `int` function to convert a value of any suitable type into an `int`.

The neighbors of a vertex are simply a list of comma separated values:

```
neighbors: { [] }
| neighbor { [$1] }
| neighbor COMMA neighbors { $1 :: $3 }
;
```

Similarly, a vertex and list of vertices are a vector position and sequence of neighbors and a comma-separated list, respectively:

```
vertex:
| OPEN vec3 COMMA OPEN neighbors CLOSE CLOSE
    { { position = $2; neighbors = set $5 } }
;

vertices:
| vertex { [$1] }
| vertex COMMA vertices { $1 :: $3 }
;
```

Note the use of the `set` function to convert any sequence into a `Set`.

Finally, a whole system is composed of the supercell dimensions (as a vector) and a list of vertices:

```
system:
| OPEN vec3 COMMA OPEN vertices CLOSE CLOSE
    { { supercell = $2; vertices = Array.of_list $5 } }
;
```

The grammatical structure of the format could have been parsed generically into a more general purpose data structure and then dissected by F# functions in the main program but this task is better suited to the optimizing compilation of grammars by a parser generator such as `fsyacc`.

12.3.4 Lexer

The lexer is very simple, handling only numbers, commas and braces.

The lexer uses the definitions of tokens from the parser:

```
{
  open Parser
}
```

A regular expression `real` handles both integers (sequences of digits) and fractional numbers (with a decimal point) and also permits a preceding minus sign for negative numbers:

```
let space = [' ' '\t' '\r' '\n']
let digit = ['0'-'9']
let real = '-'? (digit+ | digit+ '.' digit*)
```

More sophisticated regular expressions could be used but these suffice for this particular example.

The lexer itself simply reduces a character stream to lexical tokens in the simplest possible way:

```
rule token = parse
| space { token lexbuf }
| real  { REAL(Lexing.lexeme lexbuf |> float) }
| ','   { COMMA }
| '{'   { OPEN }
| '}'   { CLOSE }
```

The first case in this lexer matches whitespace using the `space` regular expression and calls the `token` rule of the lexer recursively in order to ignore the whitespace and return the next valid token. The second case matches the `real` regular expression and converts the matched string into a `float` and stores it in a `Real` token. The three final cases convert characters into tokens.

12.3.5 Main program

The main program in "nth.fs" provides an `Nth` module that parses an input file and computes the neighbors of a vertex:

```
#light
```

The following file is used for input:

```
let filename =
   __SOURCE_DIRECTORY__ + @"\cfg-100k-aSi.txt"
```

The parser is used to convert the input file into the definition of a model structure:

```
let system =
   use stream = System.IO.File.OpenRead filename
   use reader = new System.IO.BinaryReader(stream)
   let lexbuf = Lexing.from_binary_reader reader
   try
     Mmaparse.system Mmalex.token lexbuf
   with
   | Parsing.Parse_error ->
       let p = Lexing.lexeme_end_p lexbuf
       eprintf "Error at line %d\n" p.Line
       exit 1
```

In the event of a grammatical error in the input file, the `Parse_error` exception is raised by the parser. Handling this exception and printing out a descriptive error message makes this program more user friendly.

The core of the program is the function that computes the nth-nearest neighbours. This functionality relies upon the definition of the model type from the `Model` module:

```
open Model
```

The core of the program is simplified by some simple auxiliary functions. The first such function displaces a vertex from the origin supercell into a different supercell:

```
let displace i j =
   { j with offset = i.offset + j.offset }
```

This `displace` function is used when looking up the neighbor `j` of a given vertex `i` in any supercell via the neighbors of its mirror image in the origin supercell, in order to displace it to the appropriate supercell.

Next, a `unions` function that computes the n-ary union of a sequence of sets:

```
let unions : (seq<Set<coord>> -> Set<coord>) =
   Seq.fold Set.union Set.empty
```

As this is a form of dynamic programming, the `nth` function is ideally suited to memoization:

```
let memoize f =
   let m = Hashtbl.create 1
   fun k ->
     match Hashtbl.tryfind m k with
```

```
    |  Some f_k -> f_k
    |  None ->
        let f_k = f k
        m.[k] <- f_k
        f_k
```

Finally, the nth function itself is almost a direct implementation of the mathematical definition of nth-nearest neighbors that returns the singleton set for the 0th-nearest neighbor of any vertex, looks up the nearest neighbours for $n = 1$ and breaks the general problem into smaller but overlapping subproblems for $n > 1$:

```
let rec nth =
  memoize (fun n ->
    memoize (fun i ->
      match n with
      |  0 -> Set.singleton i
      |  1 ->
          system.vertices.[i.index].neighbors
          |> Set.map (displace i)
      |  n ->
          let s1, s2 = nth (n-1) i, nth (n-2) i
          unions (Seq.map (nth 1) s1) - s2 - s1))
```

This example clearly demonstrates the brevity of the F# programming language, with a complete lexer, parser and computational algorithm fitting in such a tiny amount of code. However, the F# programming language is unique in combining this brevity with incredible expressiveness and performance. This is most easily demonstrated by including a real-time interactive visualization of the results that is updated as they are computed.

12.3.6   Visualization

The simplest way to handle this and many other forms of vizualization is using the *F# for Visualization* library from Flying Frog Consultancy.

The following preamble is required to reference libraries and open relevant namespaces:

```
#I "C:\Program Files\Reference Assemblies\Microsoft\
Framework\v3.0"
#R "C:\Program Files\FlyingFrog\FlyingFrog.Graphics.dll"
open System.Windows.Media
open System.Windows.Media.Media3D
open FlyingFrog.Graphics3D
```

The following line of code spawns a concurrent visualization that may be transparently updated, with all thread-related issues handled automatically by the F# for Visualization library:

```
let viz = Show(Group [])
```

This visualization will represent each of the nth-nearest neighbors as a tiny sphere:

```
let sphere = Shape(Sphere.make 0, Brushes.White)
```

The argument to the Sphere.make function dictates the level of tesselation (as described in chapter 7) and, in this case, a coarse tesselation is used as this is satisfactory for displaying large numbers of small spheres.

Any vertex may be chosen as the origin and, in this case, the vertex with zero index in the origin supercell is used:

```
let i = {offset=vector [0.0; 0.0; 0.0]; index=0}
```

The following position function computes the actual 3D coordinates of a vertex from its index and supercell offset:

```
let position c =
  system.vertices.[c.index].position +
    system.supercell .* c.offset
```

Note the use of the . * operator for element-wise multiplication of a pair of vectors.

The following plot function computes the nth-nearest neighbours of the vertex i, builds a scene graph with a sphere at the position of each vertex and sets the scene graph as the current scene for the visualization that was spawned using the Scene property of the viz object:

```
let plot n =
  let ns = nth n i
  let at c =
    let r = position c - position i
    let r = 0.02 * r
    let move = TranslateTransform3D(r.[0], r.[1], r.[2])
    let scale = ScaleTransform3D(0.05, 0.05, 0.05)
    Transform(scale.Value * move.Value, sphere)
  viz.Scene <- Group(Seq.map at ns)
```

An animation of the nth-nearest neighbors up to $n = 40$ may then be visualized using:

```
for n in 0 .. 40 do
  plot n
```

Finally, the following function is used to join the GUI thread with the main thread to keep the whole application alive until the visualization is closed:

```
FlyingFrog.Graphics.Run()
```

This is all of the code required to produce a high-performance real-time interactive 3D visualization of the results, illustrated in figure 12.4.

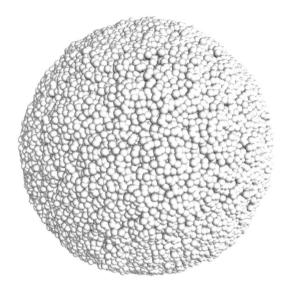

Figure 12.4   The 50th-nearest neighbour shell from a 10^5-atom model of amorphous silicon [1] rendered using the *F# for Visualization* library.

12.4 LOGISTIC MAP

The program developed in this section combines two useful concepts found in scientific computing:

- Chaotic behaviour from a simple system.

- Simple visualization.

The logistic map is a polynomial mapping that illustrates how chaotic behaviour can arise from very simple dynamics. The logistic map is described by a single mathematical definition:

$$x_{n+1} = rx_n(1 - x_n)$$

where x_n is the population at time n, r is a positive constant.
This simple relationship captures two effects from population biology:

- reproduction where the population will increase at a rate proportional to the current population when the population size is small.

- starvation (density-dependent mortality) where the growth rate will decrease at a rate proportional to the value obtained by taking the theoretical "carrying capacity" of the environment less the current population.

The following program simulates the logistic map and visualizes the results as simply as possible, using only Windows Forms. Although it is a complete Windows Forms application, this program can be run directly from an F# interactive session without any batch compilation.

The program is written in the #light syntax and opens two namespaces to simplify the subsequent code:

```
> #light;;
> open System.Drawing;;
> open System.Windows.Forms;;
```

A generic pixel format is used for the form even though the output of this program is monochrome:

```
> let format = Imaging.PixelFormat.Format24bppRgb;;
val format : Imaging.PixelFormat
```

The bitmap_of function creates a new bitmap with the dimensions of the given rectangle r and applies the given function f to fill the bitmap:

```
> let bitmap_of draw (r : Rectangle) =
    let bitmap = new Bitmap(r.Width, r.Height, format)
    draw bitmap
    bitmap;;
val bitmap_of : (Bitmap -> unit) -> Rectangle -> Bitmap
```

The resize function is an event-driven callback that replaces the given bitmap reference b, filling it in using the higher-order bitmap_of function and redraws the given form w:

```
> let resize f (b : Bitmap ref) (w : #Form) _ =
    b := bitmap_of f w.ClientRectangle
    w.Invalidate();;
val resize :
  (Bitmap -> unit) -> Bitmap ref -> #Form -> 'b -> unit
```

The paint function is an event-driven callback that draws the bitmap b onto a form using its event argument e:

```
> let paint (b : Bitmap ref) (v : #Form)
      (e : PaintEventArgs) =
    let r = e.ClipRectangle
    e.Graphics.DrawImage(!b, r, r, GraphicsUnit.Pixel);;
val paint :
  Bitmap ref -> #Form -> PaintEventArgs -> unit
```

The make_raster function creates a form and a bitmap and registers the resize, paint and key down callbacks:

```
> let make_raster title f =
    let form = new Form(Text=title, Visible=true)
    let bitmap = ref (bitmap_of f form.ClientRectangle)
```

```
  form.Resize.Add(resize f bitmap form)
  form.Paint.Add(paint bitmap form)
  form.KeyDown.Add(fun e ->
    if e.KeyCode = Keys.Escape then form.Close())
  form;;
val make_raster : string -> (Bitmap -> unit) -> Form
```

The function that determines the evolution of a population from one time step to the next may be written:

```
> let f r x =
    r * x * (1.0 - x);;
val f : float -> float -> float
```

The draw function is responsible for filling in the pixels of the bitmap and will be invoked when it is passed as the argument to the bitmap_of function:

```
> let draw (bitmap : Bitmap) =
    let w, h = bitmap.Width, bitmap.Height
    for j=0 to h-1 do
      for i=0 to w-1 do
        bitmap.SetPixel(i, j, Color.White)
    for i=0 to w-1 do
      let r = 2.4 + float i / float w * 1.6
      for j=0 to h-1 do
        let y = (float j + 0.5) / float h
        let y = nest 1000 (f r) y
        let j = int_of_float ((1.0 - y) * float h)
        bitmap.SetPixel(i, j, Color.Black);;
val draw : Bitmap -> unit
```

This draw function uses the nest combinator from section 6.1.1.

Finally, a Windows form visualizing the logistic map may be created using the make_raster function:

```
> let form = make_raster "Logistic map" draw;;
val form : Form
```

The result is illustrated in figure 12.5.

The functions that make up this program could have been simplified slightly by defining the draw function first and calling it explicitly from the functions relating to the paint callback. However, this would prevent reuse of the code. Specifically, parameterizing the bitmap_of function used to fill the bitmap over the draw function that does the filling allows different draw functions to be used.

12.5 REAL-TIME PARTICLE DYNAMICS

The program developed in this section combines two useful concepts found in scientific computing:

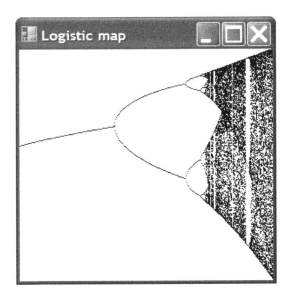

Figure 12.5   Chaotic behaviour of the logistic map.

- Simulation of particle dynamics.

- High-performance real-time interactive visualization that runs concurrently with the main computation.

The following program simulates dynamics of a system of non-interacting particles and visualizes the results interactively and in real time using the *F# for Visualization* library. Although this is a complete GUI application, this program can be run directly from an F# interactive session without any batch compilation.

The program begins by referencing the *F# for Visualization* library:

```
#light
#I "C:\Program Files\Reference Assemblies\Microsoft\
Framework\v3.0"
#I @"C:\Program Files\FlyingFrog"
#R @"FlyingFrog.Graphics.dll"
```

The following namespaces are opened in order to simplify the subsequent code:

```
open System.Windows.Media
open System.Windows.Media.Media3D
open FlyingFrog.Graphics3D
```

The particles will bounce around on a 3D surface quantified by the following function:

```
let f x z =
  let r = 5.0 * sqrt(x*x + z*z)
```

```
let sinc = function
  | 0.0 -> 1.0
  | x -> sin x / x
sinc r + 0.01 * sin x + 3e-3*r*r - 1.12
```

The position and velocity of each particle are encapsulated in the following record type:

```
type particle = { p: Vector3D; v: Vector3D }
```

All of the particles have the same radius:

```
let size = 0.015
```

The following function creates a new randomly-placed particle:

```
let rand = new System.Random()
let spawn _ =
  let f() = rand.NextDouble() * 0.02 - 0.01
  { p = Vector3D(f(), 3.0, f());
    v = Vector3D(0.0, 0.0, 0.0) }
```

The current state of the particle system is represented by an array:

```
let state = Array.init 100 spawn
```

Gravity is quantified by the following vector:

```
let gravity = Vector3D(0.0, -1.0, 0.0)
```

Air resistance is quantified by the following number:

```
let air_loss = 0.9
```

Energy lost by collision with the floor is quantified by the following number:

```
let bounce_loss = 0.99
```

When the dynamics of a particle are integrated over a time span that includes a collision, the simulation will be subdivided down to time spans below the following value:

```
let max_dt = 1e-3
```

The following numbers are the minimum velocity and minimum height, below which particles will be respawned to keep the simulation interesting:

```
let min_v2, min_h = 1e-3, 1e-3
```

The following function integrates the equations of motion for a single particle a for a time step of length dt, including acceleration due to gravity and damping of the velocity:

```
let rec update dt ({p=p; v=v} as a) =
  let p = p + dt * v + 0.5 * dt*dt * gravity
```

```
let v = air_loss ** dt * (v + dt * gravity)
```

Note how the record fields p and v are extracted in a pattern match along with the record a itself using a named subpattern (see section 1.4.2.2).

The height of particle above the floor at the x, z coordinate of the particle is used to test for impact:

```
let height = p.Y - f p.X p.Z - size
if v.LengthSquared < min_v2 && height < min_h then
  spawn()
else
  if height >= 0.0 then { p = p; v = v } else
    if dt > max_dt then
      let f = update (dt / 2.0)
      f(f a)
    else
      let n = surface_normal f (p.X, p.Z)
      let d = Media3D.Vector3D.DotProduct(n, v)
      if d > 0.0 then {p=p; v=v} else
        { p = p; v = bounce_loss * (v - 2.0 * d * n) }
```

In particular, if the time interval dt spanning a collision is greater than max_dt then the time interval is recursively bisected. This is a seemingly-trivial addition that is very easily implemented in F#. However, this part of the algorithm is of critical importance. The time integrator is exact to within numerical error for quadratic behaviour and very accurate for the slightly-damped ballistic behaviour of a flying particle but wildly inaccurate for collisions, when the velocity of the particle is instantaneously replaced at some interim moment in time. By recursively bisecting time intervals that span collisions, the collision is limited to a single short time interval where the error in time integration is much smaller. Real scientific simulations of particle systems often use similar tricks to greatly improve the accuracy of the simulation with minimal adverse effect on performance [8].

The F# for Visualization library is specifically designed to handle real-time simulations and, consequently includes an accurate timer. This timer is represented by a curried function delta_timer. Applying the first argument () to delta_timer returns a function that gives the time since it was last invoked. Consequently, this delta_timer function can be used to create many independent timers. In this case, we need only one:

```
let dt = FlyingFrog.Timer.delta_timer()
```

The following function calculates a transformation matrix that will be used to scale a scene graph representing a unit sphere to the size of a particle and then translate it to the particle's position in 3D space:

```
let particle p =
  ScaleTransform3D(size, size, size).Value *
    TranslateTransform3D(p.p).Value
```

The Show3D constructor from the F# for Visualization library spawns a visualization that runs concurrently with the current thread. In this case, the scene is initially empty:

```
let g = Show(Group [])
```

The following scene graph is a smooth icosahedron that will be used to visualize each particle in the simulation:

```
let sphere =
  let color = System.Windows.Media.Brushes.White
  Shape(Sphere.make 0, color)
```

The following scene graph is a tesselation of a relevant part of the floor that will be used in the visualization:

```
let floor =
  let s = -3.0, 3.0
  let color = System.Windows.Media.Brushes.Red
  Shape(plot3d 128 f s s, color)
```

The following function simulates the particle dynamics by looping indefinitely and repeatedly replacing the scene graph that is being rendered by the concurrent visualization:

```
let simulate() =
  while true do
    System.Threading.Thread.Sleep(30)
    let particles =
      [|for p in state ->
          Transform(particle p, sphere)|]
    g.Scene <- Group[Group particles; floor]
    let dt = dt()
    for i=0 to state.Length-1 do
      state.[i] <- update dt state.[i]
```

The simulation is run in a new background thread:

```
let simulator =
  let thread = new System.Threading.Thread(simulate)
  thread.IsBackground <- true
  thread.Start()
```

As the simulation thread is a background thread and the visualization thread is (by default) a foreground thread, the program will run until the visualization is closed by the user, at which point the foreground visualization thread will die and the background simulation thread will be terminated.

Finally, the following function is used to join the GUI thread with the main thread to keep the whole application alive until the visualization is closed:

```
FlyingFrog.Graphics.Run()
```

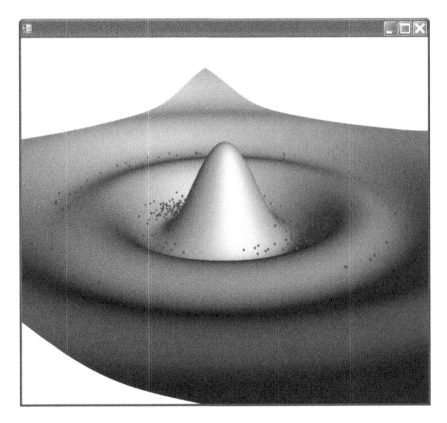

Figure 12.6 Real-time interactive simulation and visualization of non-interacting particles bouncing on a 3D surface.

The result is illustrated in figure 12.6.

Appendix A

Troubleshooting

This appendix details many of the problems commonly encountered by new F# programmers.

A.1 VALUE RESTRICTION

One problem encountered when using an F# interactive session is the value restriction, which requires types to be completely defined.

The simplest example impeded by the value restriction is the definition of a reference to an empty list:

```
> let x = ref [];;
  ----^^stdin(2,4): error: FS0030: Value restriction.
Type inference has inferred the signature
      val x : '_a list ref
Either define 'x' as a simple data term, make it a
```

*

function, or add a type constraint to instantiate the
type parameters.

The empty list has the well-defined type `'a list` but a reference to the empty
list does not have a well-defined type because the empty list might be replaced by
a non-empty list, in which case the elements must be of a specific type, the type of
which has not been defined. In this case, the value restriction protects the programmer
from accidentally setting the reference to an `int list` in one part of the program
but a `float list` in another part of the program.

This problem can always be addressed by adding a type annotation to constrain
the type of the value. In this case, we might know that the list will contain `int`
values, so we annotate accordingly to make the F# interactive session happy:

```
> let x : int list ref = ref [];;
val x : int list ref
```

The problem occurs mainly when executing parts of a program being developed
in an F# interactive session.

A.2 MUTABLE ARRAY CONTENTS

If an array is initialized from a reference value using the `Array.create` function
then the reference is shared between every element of the array. For example, the
following creates an array containing 3 elements, all of which are the same reference
to an integer:

```
> let a = Array.create 3 (ref 0);;
val a : int ref array
```

Assigning the reference from any element causes all elements to be assigned:

```
> a.[0] := 7;;
> a;;
val it : int ref array =
  [|{contents = 7}; {contents = 7}; {contents = 7}|]
```

This problem is most easily solved by using the `Array.init` function to create
the array, specifying a function that returns a new reference at each invocation:

```
> let a = Array.init 3 (fun _ -> ref 0);;
val a : int ref array
```

or using a comprehension:

```
> let a =
    [|for i in 0 .. 2 ->
        ref 0|];;
```

Every array element now contains its own mutable value:

```
> a.[0] := 7; a;;
```

```
val it : int ref array =
  [|{contents = 7}; {contents = 0}; {contents = 0}|]
```

This caveat applies to any instantiation of an array with mutable elements, including arrays of hash tables.

A.3 NEGATIVE LITERALS

As a special case, floating point literals that are arguments do not need to be bracketed. For example, `f (-1.0) (-2.0)` may be written `f -1.0 -2.0`.

Although very useful, this syntax is currently somewhat fragile. In particular, the positions of spaces are important. For example, the following is interpreted incorrectly as a function application rather than a subtraction:

```
> 1.0-2.0;;
stdin(3,0): error: FS0003: This is not a function and
cannot be applied. Did you forget a ';' or 'in'?
```

F# understood that to mean `1.0(-2.0)`, i.e. applying `-2.0` as an argument. This example must therefore be spaced out:

```
> 1.0 - 2.0;;
val it : float = -1.0
```

Consequently, it is good style to add spaces between operators when possible, in order to disambiguate them.

A.4 ACCIDENTAL CAPTURE

Closures are local function definitions that can capture local variables. Captured variables are stored inside the closure by the F# compiler and are referred to as the *environment* of the closure. This is a powerful form of automated abstraction and is one of the cornerstones of functional programming.

This form of abstraction has an important caveat that F# programmers should be aware of. The environment of a closure can accidentally capture important variables such as file handles and keep them alive for longer than expected. This should not be a problem for well written programs as external resources such as file handles should always be closed explicitly via the use construct but programmers often succumb to the temptation of abusing the garbage collector for the release of external resources. This can cause leaks if closures capture references that keep the resources alive.

A.5 LOCAL AND NON-LOCAL VARIABLE DEFINITIONS

Although similar in appearance, the let keyword is used in two different ways and this can be invasive when using the traditional syntax. Specifically, non-nested (outermost):

```
let ... = ...
```

constructs make new definitions in the current namespace whereas nested:

```
let ... = ... in
...
```

constructs make local definitions.

The difference between nested and non-nested definitions can sometimes be confusing. For example, the following is valid F# code (written in traditional syntax for clarity) that defines a variable a:

```
> let a =
    let b = 4 in
    b * b;;
val a : int = 16
```

In contrast, the following tries to make a non-local definition for a within the nested expression for b, which is invalid:

```
> let b = 4 in
  let a = b * b;;
Syntax error
```

This is one of the trivial sytactic mistakes often made when learning F#.

A.6 MERGING LINES

The #light syntax option circumvents some of the problems with the traditional syntax but also introduces new problems. For example, the following is intended to be read as a pair of separate lines:

```
let links = getWords >> List.filter ((=) "href")
google |> links
```

Attempting to evaluate these two lines at once (by selecting them both and pressing ALT+ENTER) gives an error because the compiler actually interprets the code as:

```
let links =
  getWords >> List.filter ((=) "href")
  google |> links
```

This ambiguity can be resolved in various ways but the simplest solution is to use the ;; token to delimit the blocks of code.

A.7 APPLICATIONS THAT DO NOT DIE

Multithreaded GUI applications sometimes suffer from the problem that a worker thread continues executing unnecessarily after all user-visible functionality has disappeared when an application is closed. This can be addressed by marking the worker thread as a background thread using the `IsBackground` property of the `Thread` object.

A.8   BEWARE OF "IT"

Evaluation of an expression rather than a definition in an interactive session yields a response of the form:

```
val it : ...
```

As you may have guessed, this actually defines a variable `it` to have the value resulting from the evaluation, i.e. `it` may be used to refer to the last result.

This can be very useful. However, it will overwrite any variable called `it` that you have defined and `it` is a tempting variable name to use when referring to an iterator.

Glossary

λ-**function**   an anonymous function.

Abstract Syntax Tree   the data structure representing a symbolic expression or program, with operators at nodes and values (typically symbols or numbers) at the leaves.

Abstract type   a type with a visible name but hidden implementation. Abstract types are created by declaring only the name of the type in a module signature, and not the complete implementation of the type as given in the module structure.

Accumulator   a variable used to build the result of a computation. The concept of an accumulator underpins the fold algorithm (introduced on page 40). For example, in the case of a function which sums the elements of a list, the accumulator is the variable which holds the cumulative sum while the algorithm is running.

ADO.NET   part of the .NET framework that deals with database programming.

Algorithm   a mathematical recipe for solving a problem. For example, Euclid's method is a well-known algorithm for finding the largest common divisor of two numbers.

Array   a flat container which provides random access to its elements in $O(1)$ time-complexity. See section 3.2.

ASP.NET   part of the .NET framework that deals with web programming.

Associative container   a container which represents a mapping from keys to values.

Asymptotic complexity   an approximation to the complexity of an algorithm, derived in the limit of infinite input complexity and, typically, as a lower or upper bound. For example, an algorithm with a complexity $f(n) = 3 + 3n + n^2$ has an asymptotic complexity $O(n^2)$. See section 3.1.

Balanced tree   a tree data structure in which the maximum variation in depth is restricted. See section 3.10.1 for a brief discussion.

Big int   an arbitrary-precision integer type. Slower to use than an ordinary `int` but can represent arbitrarily-large numbers.

Binary tree   a tree data structure in which all non-leaf nodes contain exactly two binary trees.

C#   an object-oriented programming language for the .NET platform.

Cache   an intermediate store used to accelerate the fetching of a subset of data.

Cache hit   the quick process of retrieving data which is already in the cache.

Cache miss   the slow process of fetching data to fill the cache when a request is made for data not already in the cache.

Cache coherent   improving the locality of memory accesses in order to improve the effectiveness of caching, e.g. by replacing random accesses with sequential accesses.

Cartesian cross product   a set-theoretic form of outer product. For example, the cartesian cross product of the set $A = \{a, b\}$ with the set $B = \{c, d, e\}$ is the set of pairs $A \times B = \{(a, c), (a, d), (a, e), (b, c), (b, d), (b, e)\}$.

class   a type of object that may encapsulate related value and type definitions.

Closure   a function or object value that may capture variables from its environment such as a partially-applied curried function.

Compile-time   while a program is being compiled.

Compiler   a program capable of transforming other programs. For example, the `fsc` compiler transforms F# source code into executable machine code.

Complexity   a quantitative indication of the growth of the computational requirements (such as time or memory) of an algorithm with respect to its input. Algorithmic complexity is described in section 3.1.

Cons   the `::` operator. When used in a pattern, `h::t` is said to *decapitate* a list, binding `h` to the first element of a list (the *head*) and `t` to a list containing the remaining elements (the *tail*). When used in an expression, `h::t` prepends the element `h` onto the list `t`. See sections 3.3 and 6.4 for example uses of the cons operator.

Container   a data structure used to store values. The values held in a data structure are known as the *elements* of the data structure. Arrays, lists and sets are examples of data structures.

Curried function   any function that returns a function as its result. See section 1.6.3.

Data structure   a scheme for organizing related pieces of information.

Decapitate   splitting a list into its first element (the *head*) and a list containing the remaining elements (the *tail*).

Delegate   a .NET construct similar to a function value in functional programming except that the type of a delegate must be explicitly declared. Delegates are not directly related to F# but may be used during interop with .NET libraries written in other languages, such as C#.

DirectX   a collection of application programming interfaces for handling tasks related to multimedia, especially game programming and video, on Microsoft platforms.

Environment   the declaration of a function value that refers to variables defined in the local scope captures those variables. The variables captured by a closure are referred to as its environment.

Exception   a programming construct which allows the flow of execution to be altered by the raising of an exception. Execution then continues at the most recently defined exception handler capable of dealing with the exception. See section 1.4.5.

Flat container   a non-hierarchical data structure representing a collection of values (elements). For example, arrays and lists.

FIFO   first-in first-out semantics for a container, where the first value inserted into a container will be the next value removed from it.

Fixed point   an int and a (possibly implicit) scaling. Used to represent real-valued numbers $x \in \mathbb{R}$ approximately, with a constant absolute error.

Float   the type of a double-precision IEEE floating-point number.

Floating point   a number representation commonly used to approximate real-valued numbers $x \in \mathbb{R}$. See section 4.1.2.

Folds   a higher-order function which applies its function argument to an accumulator and each element of a container. Introduced on page 40.

Function   a mapping from input values to output values which may be described implicitly as an algorithm or explicitly, e.g. by a pattern match.

Functional programming   a style of programming that emphasizes the use of functions.

Functional language   a programming language that provides first-class lexical closures.

Functor   a construct that maps modules to modules. Used extensively in the ML family of languages but not directly supported by F#, which uses object-orientation to solve the same problem.

Garbage collection   the process of identifying data that are no longer accessible to a running program, disposing of them and reclaiming the resources they required.

Generic programming   the use of polymorphic functions and types.

Graph   a data structure composed of vertices, and edges that link pairs of vertices. Used to model networks such as links between websites and metabolic pathways in cells.

Hash   a constant-sized datum computed from an arbitrarily complicated value. Unequal hashes implies inequality but equal hashes do not imply equality. Most often used to accelerate searching by culling values with different hashes from the search, such as when searching for a key in a hash table.

Hash table   a data structure providing fast random access to its elements via their associated hash values. See section 3.5.

Head   the element at the front of a list. See section 3.3.

Higher-order function   any function which accepts another function as an argument. For example, f is a higher-order function in the definition $f(g, x) = g(g(x))$ because g must be a function (as g is applied to x and then to $g(x)$). See section 1.6.4.

Heterogeneous container   a data structure capable of storing several elements of different types. See section 3.9.

Homogeneous container   a data structure (container) capable of storing several elements of the same type. For example, an array of integers is a homogeneous container because an array is a data structure containing elements of a single type, in this case integers.

Imperative programming   a style of programming in which the result of a computation is generated by statements which act by way of side-effects, as opposed to functional programming.

Iteration   a homonym with different meanings in different contexts and disciplines. In the context of numerical algorithms, an "iterative algorithm" means an algorithm designed to produce an approximate result by progressively converging on the solution. More generally, the word iterative is often used to describe repetitive algorithms, where a single repeat is known as an *iteration*.

Impure functional language   a language, such as F#, that provides both functional and imperative programming constructs.

Int   a type which exactly represents a contiguous subset of the integers $\mathbb{Z}$. See section 4.1.1.

Interactive mode   a way of executing F# code piece by piece, in contrast to compiling an F# program into an executable.

IO   input and output operations, such as printing to the screen or reading from disc.

Lazy evaluation   an evaluation strategy where expressions are only evaluated when their result in needed. Antonym of *strict*. F# provides several lazy constructs, most notably the Seq module and the lazy function.

Leaf   in the context of tree data structures, a leaf node is a node containing no remaining trees.

Lex   converting a character stream into a token stream. For example, recognising the keywords in a language before parsing them.

LIFO   last-in first-out semantics for a container, where the last value inserted into a container will be the next value removed from it. A *stack* is a LIFO container.

Linked List   see list.

List a flat container providing prepend and decapitation in $O(1)$ time-complexity. In F#, these are performed by the : : operator, known as the *cons* operator. A list is traversed by repeated decapitation. See section 3.3.

Maps either a container or a higher-order function:

- A data structure implementing a container which allows key-values pairs to be inserted and keys to be subsequently mapped onto their corresponding values. See sections 3.5 and 3.6.

- A higher-order function $/map\ f\ \{l_0, \ldots, l_{n-1}\} \rightarrow \{f(l_0), \ldots, f(l_{n-1})\}$ which acts upon a container of elements to create a new container, the elements of which are the result of applying the given function f to each element l_i in the given container. Sometimes known as *inner map*.

Module a construct which encapsulates definitions in a *structure* and, optionally, allow the externally-visible portion of the definitions to be restricted to a subset of the definitions by way of a *signature*. See section 2.3.

Module signature a module interface, declaring the types, exceptions, variables and functions which are to be accessible to code outside a module using the signature.

Module structure the body of a module, containing definitions of the constituent types, exceptions, variables and functions which make up the module.

Monomorphic a single, possibly not-yet-known, type.

Mutable can be altered.

Native code the result of compiling a program into the machine language (machine code) understood natively by the CPU.

.NET the core of Microsoft's next generation platform, combining seamless inter-operability between programs written in different languages with a comprehensive suite of libraries.

Object an instantiation of a class as a value.

Object-oriented programming the creation of objects, which encapsulate functions and data, at run-time. In particular, the use of inheritance to specify relationships between types of object.

Parse the act of understanding something formally. Parsing often refers to the recognition of grammatical constructs. See section 5.5.2.

Partial specialisation the specialisation of a program or function to part of its input data. For example, given a function to compute x^n for any given floating-point number x and integer n, generating a function to compute x^3 for any floating-point number x is partial specialising the original to $n = 3$.

Pattern matching a programming language construct provided by F# (and other members of the ML family) that allows a value to be compared against a sequence of patterns in order to determine the course of action. In particular, pattern matching facilitates the destructuring of values of a compound type into its subvalues, such as extracting branches from a tree. See section 1.4.2.

Persistence   the ability to reuse old data structures without having to worry about undoing state changes and unwanted interactions. An advantage of functional programming.

Platform   a CPU architecture (e.g. ARM, MIPS, AMD, Intel) and operating system (e.g. IRIX, Linux, Mac OS X, Windows XP).

Polymorphic   one of any type. In particular, polymorphic functions are generic over the types of at least one of their arguments. Variant types can be generic over polymorphic type-arguments.

Primitive operation   a low-level function or operation, used to formulate the time-complexity of an algorithm. See section 3.1.1.

Purely functional language   a language (like Haskell) that prohibits the use of side effects.

Record   a product type with named fields. For example, a record of type: `{ x: float; y: float }` can have a value $\{x=1.0; y=2.0\}$.

Reference counting   a slow and inaccurate form of garbage collection where the number of references to each value is stored and values are deallocated only if their reference count reaches zero. This fails to account for cyclic data structures where reference counts never reach zero even though the structure is unreachable and should be deallocated.

Regular Expression   a form of pattern matching.

Regexp   common abbreviation of *regular expression*.

Root   in the context data structures, the root is the origin of the data structure, from which all other portions may be accessed.

Run-time   while a program is being executed.

Search tree   a tree where the order of subtrees reflects an ordering of the values in the tree, allowing subtrees to be culled when searching for a particular value or range of values. In F#, the `Set` and `Map` modules provide implementations of sets and maps based upon balanced binary search trees. Section 3.13 describes the use of unbalanced search trees in computational science.

Side-effect   any result of an expression apart from the value which the expression returns, e.g. altering a mutable variable or performing IO.

Signature   see *Module signature*.

Source code   the initial, manually-entered form of a program.

Stack   either:

- a data structure providing LIFO access.

- a limited resource provided by the operating system designed for the storage of local variables during function recursion. Overuse of the system stack during recursion is made possible by tail recursion in F#.

Static typing   completely type checking at compile-time such that no type checking is required at run-time.

Strict   expressions are evaluated immediately and the resulting value is used in subsequent operations. Antonym of *lazy*.

Structure   see *Module structure*.

struct   an unboxed representation in .NET, similar to a class.

Tail   the remainder of a list without its front element.

Tail call   a function call whose result is returned immediately (with subsequence computation or exception handling).

Tail recursion   a recursive function where the recursive calls are tail calls.

Time-complexity   complexity of the time taken to execute an algorithm, specified as the number of times a set of primitive operations are performed.

Thread   a construct used to represent a computation such that threads may be executed concurrently to achieve parallelism.

Tree   a recursive data structure represented by nodes which may contain further trees. The root node is the node from which all others may be reached. Leaf nodes are those which contain no further trees. Trees are traversed by examining the child nodes of the current node recursively.

Tuple   a product type with unnamed elements. Tuples represent elements in the set of the cartesian cross product of the sets of element types in the tuple. For example, the 2-tuple of floating-point numbers (x, y) of the type `float * float` is typically used to represent the set $\mathbb{R} \times \mathbb{R}$.

Type   the set of possible values of a variable, function argument or result, or the mapping between argument and result types of a function.

Variant type   explicitly listed sets of possible values. See section 1.4.1.3.

Visual Basic   an object-oriented .NET programming language that focuses on ease of use.

Bibliography

1. Gerard T. Barkema and Normand Mousseau. High-quality continuous random networks. *Phys. Rev. B*, 62:4985, 2000.

2. T. A. Brody, J. Flores, J. B. French, P. A. Mello, A. Pandey, and S. S. M. Wong. Random-matrix physics - spectrum and strength fluctuations. *Rev. Mod. Phys.*, 53:385, 1981.

3. Thomas H. Cormen, Charles E. Leiserson, Ronald L. Rivest, and Clifford Stein. *Introduction to algorithms*. MIT Press, Cambridge, MA, USA., 2001.

4. I. Daubechies. Orthonormal bases of compactly supported wavelets. *Comm. Pure Appl. Math.*, 41:909, 1988.

5. S. R. Elliott. *The physics and chemistry of solids*. John Wiley & sons, New York, USA, 2000.

6. T. Fischbacher, T. Klose, and J. Plefka. Planar plane-wave matrix theory at the four loop order: Integrability without bmn scaling. *J. High Energy Phys.*, 2005(02), 2005.

7. D. S. Franzblau. Computation of ring statistics for network models of solids. *Phys. Rev. B*, 44(10):4925–4930, 1991.

8. D. Frenkel and B. Smit. *Understanding Molecular Simulation from Algorithms to Applications*. Academic Press, New York, USA, 1996.

9. Matteo Frigo and Steven G. Johnson. FFTW: An adaptive software architecture for the FFT. In *Proc. IEEE Intl. Conf. on Acoustics, Speech, and Signal Processing*, volume 3, pages 1381–1384, Seattle, WA, May 1998.

10. Stephen Gasiorowicz. *Quantum physics*. John Wiley and Sons, London, England, 2003.

11. J. D. Harrop. *OCaml for Scientists*. Flying Frog Consultancy, Cambridge, UK, 2005.

12. Jon D. Harrop. Balanced binary search trees. *The F#.NET Journal*, 2007.

13. Jon D. Harrop. Combinator heaven. *The F#.NET Journal*, 2007.

14. Jon D. Harrop. Language-oriented programming: The term-level interpreter. *The F#.NET Journal*, 2007.

15. Jon D. Harrop. Quick introduction to web services and databases. *The F#.NET Journal*, 2007.

16. Jon D. Harrop. Tail recursion. *The F#.NET Journal*, 2007.

17. Donald E. Knuth. *The Art of Computer Programming*. Addison Wesley, Boston, MA, USA, 1997.

18. C. J. Lanczos. A precision approximation of the gamma function. *J. SIAM Numer. Anal. Ser. B*, 1:86–96, 1964.

19. P. A. Lee and T. V. Ramakrishnan. Disordered electronic systems. *Rev. Mod. Phys.*, 57:287, 1985.

20. S. Mallat. Multiresolution approximations and wavelet orthonormal bases of $L^2(\mathbb{R})$. *Transactions of the American Mathematical Society*, 315(1):69–87, 1989.

21. S. G. Mallat. A theory for multiresolution signal decomposition: the wavelet representation. *IEEE Trans. on Patt. Anal. and Mach. Intel.*, 11(7):674–693, 1989.

22. Robert Pickering. *Foundations of F#*. APress, Berkeley, CA, USA, 2007.

23. W. Rankin and J. Board. A portable distributed implementation of the parallel multipole tree algorithm. In *IEEE Symposium on High Performance Distributed Computing*, pages 17–22, Los Alamitos, 1995. IEEE Computer Society Press.

24. Jonathan Shewchuck. Adaptive precision floating-point arithmetic and fast robust geometric predicates. *Discrete & Computational Geometry*, 18(3):305–363, 1997.

25. Don Syme, Adam Granicz, and Antonio Cisternino. *Expert F#*. APress, Berkeley, CA, USA, 2007.

INDEX

Printed and bound by CPI Group (UK) Ltd, Croydon, CR0 4YY

27/10/2024

14580134-0002